Surviving Climate Anxiety

Surviving Climate Anxiety

A Guide to Coping, Healing, and Thriving

Dr. Thomas Doherty

LITTLE,
BROWN
SPARK

New York Boston London

Little, Brown Spark
Hachette Book Group
1290 Avenue of the Americas, New York, NY 10104
littlebrownspark.com

First Edition: October 2025

Little, Brown Spark is an imprint of Little, Brown and Company, a division of Hachette Book Group, Inc. The Little, Brown Spark name and logo are trademarks of Hachette Book Group, Inc.

The publisher is not responsible for websites (or their content) that are not owned by the publisher.

The Hachette Speakers Bureau provides a wide range of authors for speaking events. To find out more, go to hachettespeakersbureau.com or email hachettespeakers@hbgusa.com.

Little, Brown and Company books may be purchased in bulk for business, educational, or promotional use. For information, please contact your local bookseller or the Hachette Book Group Special Markets Department at special.markets@hbgusa.com.

ISBN 9780316572781 (hc) / 9780316601092 (international tpb)
Library of Congress Control Number: 2025941795

Printing 1, 2025

LSC-C

Printed in the United States of America

It is a characteristic of wisdom not to do desperate things.
—HENRY DAVID THOREAU

TABLE OF CONTENTS

INTRODUCTION

The Climate Crisis Is Personal

The environmental anguish of the Earth has entered our lives as a radical transformation of human identity. The needs of the planet and the needs of the person have become one, and together they have begun to act upon the central institutions of our society with a force that is profoundly subversive, but which carries within it the promise of cultural renewal.

—Theodore Roszak[1]

I'm leading a workshop. At the very start I invite participants to take part in a brief exercise. I have two props, two small inflatable balls. One is a standard blue globe showing the continents and the oceans, and the other is a bright red flaming ball of the same size. I say that the blue globe represents Earth with all of its beauty and diversity, and all that we love and care about. The flaming globe represents all of the things that we are concerned about, the weight of all of the problems and uncertainties we are carrying. Several attendees take turns coming up to the front of the room. I have them hold the globes in their hands and invite them to share what they are thinking and feeling. They look down at the globes and consider, and they speak of what they are afraid of or angry about, what they cherish, their dreams and their sadness. One person slumps their shoulders as if the globes are truly heavy. Another becomes tearful as they consider the future of their children. Another hugs the blue globe to their breast as if to protect it. Another holds the globes up boldly and describes the positive vision they are working toward.

Each day you open your eyes to the world. But rarely do you have an opportunity to stop, take a breath, and gather your thoughts and feelings about the momentous events you are taking part in.

I've facilitated the two-globes exercise in classrooms, at public meetings, in therapy sessions, at businesses, on outdoor trips, and around the world— in the United States, New Zealand, South Korea, and Brazil. It works even when we don't share a common language. People get what I intend when I hold up the globes. Many times they line up, excited for a chance to share their story. It is especially poignant when children take up the globes in their small hands and tell their stories, and it's inspiring to know they are among adults who care and who are listening.

Today, out-of-control climate breakdown is the greatest public health threat humans face in the twenty-first century, or certainly a force multiplier for other threats. We're experiencing twin emergencies of a magnitude we have never seen. The first are physical harms to life caused by fire, heat, storms, and flooding. The other is a mental health crisis directly resulting from disasters, their indirect ripple effects, and the emotional toll of anxiety, loss, and depression. Anguish about a despoiled environment is not new, nor is a sense of life's precarity, but the stakes *feel* a lot higher now. And it's harder to escape.

Back in 2007, I was the lone clinical psychologist serving on the newly minted American Psychological Association's Task Force on Global Climate Change. My role was to sort out the psychological impacts of climate change. I realized that the impact of vicarious suffering could be severe. Of course, anyone in the direct path of the unnatural disasters caused by climate breakdown like unprecedented fire or floods would experience trauma. But people reading about it, worrying about it, anticipating it, would also be traumatized. Back in 2009, this was all discussed in a future tense, as speculative. But that future, one that we foresaw nearly two decades ago, has caught up to us.

When I published the groundbreaking paper "The Psychological Impacts of Global Climate Change" in *American Psychologist*, I predicted a perfect storm brewing for a mental health crisis of global proportions.[2] And here we are. This book gathers in one place all I have learned about coping since then; it is a program you can use to know yourself better, feel better, and be better at taking action that is appropriate and meaningful to you. The idea that we can thrive emotionally in these times is so revolutionary that it can be life-changing.

The term "eco-anxiety" is not new. It was first used in relation to people's worries about endocrine-disrupting chemicals in our food chain and how this affected human fertility. As evidence for rapid warming of the planet became clear and climate disasters mounted, "eco-anxiety" became synonymous with "climate anxiety."

Feeling anxious and apprehensive about the state of the planet is a normal and healthy emotion that manifests in your mind and body. Anxiety serves a specific role in your survival and functioning. It is a signal to investigate a potential threat.

Unfortunately, continuing crises and lack of a concerted international response keeps people trapped in a cycle of new threats and disasters. Moreover, the anxiety-provoking nature of all scientific reports and garish news has *itself* become a major health problem. You absorb the stress in your body and have to fend for yourself.

But you are not alone.

This book is a user's guide for coping with the mental health stresses of global climate change and other environmental issues. It is not a catalogue of the extreme weather disasters that await you, a road map to post-fossil-fuel energy transition, or a survey of the many actions you take in your daily life to be a good ecological citizen of Earth. Those topics are important, and many books have been written about them that offer much that is valuable and useful. This book is different. It is about coping and possibly thriving in the face of climate change from a psychological perspective.

You have a responsibility to cope with environmental anxiety because you have a responsibility to show up in your life and, ideally, be the best person you can. That includes evaluating ecological threats, taking action as you can, and otherwise letting go of the fear.

EXPANDING YOUR CAPACITY

To survive climate anxiety requires you to expand your capacity to take in more: more knowledge, more experience, more feelings, more kinds of relationships. The key lesson is that you need to expand in two ways—for the bad things and the good things. In his 2019 book *The Uninhabitable Earth*, journalist David Wallace-Wells writes: "It is worse, much worse, than you think."[3] I agree. Opening yourself to climate change and the plight of the natural world is opening yourself to wounds and threats that sometimes you'd rather not know about. No matter what issue, if you dig in honestly you're likely to find it more complicated, more dire, more tragic, more unjust than you anticipated. But the truth is, while the situation is worse than you know, *it is also better than you realize*. If you dig in, again honestly, you will find objectively good things. The vast majority of people around the world care about nature and want to take action on issues like climate breakdown. In whatever place you choose to explore, you will find good people, with positive values, making inspiring and creative efforts. To echo Rachel Carson, if you take time to contemplate the beauty of nature, you will experience times of connection, love, and awe, and a "sense of wonder so indestructible that it would last throughout life."[4]

Expanding your capacity is not easy, but I'll show you ways to do it. Opening yourself to the pain of ecological crises—with expert guidance and tools—also opens you to learning more about what's important to you, what you value, and how you express your relationship to the natural world, of which you are a part. It also connects you with inspiring efforts of people like yourself, all around the world, who are creating positive change. In a way, this book takes up the challenge that ecological truth-tellers like Wallace-Wells pose. Okay, it's worse than we think, worse than we know. But there are ways to be healthy and live fully with this knowledge. Despite the dire circumstances of the era we live in, there is a future and a path forward for personal growth. I do believe that we can cope and flourish in these times. Projects like the podcast I founded with my colleague Panu Pihkala, a Finnish climate emotions researcher, *Climate Change and Happiness*, tap into the positive changes happening all around us in direct response to the climate crisis. The advice I give to my clients and students is "Prepare to be blindsided, and remain open to the possibility that you will positively be surprised."

You don't need to solve climate change, or other environmental problems, in order to cope with them. But you can and should do your own personal work to show up and be your best self during the crisis, find ways to adapt to the changes you must endure, and then put your energy toward mitigating the problem and preventing needless suffering on the part of yourself and people and places you value, and on the part of other people and beings, now and in the future.

You will grow. You will learn that what we call "climate change" or the "biodiversity crisis" is a complex net of realities, intractable issues, and local emergencies that cannot be solved in any simple, linear way. Can you engage with them? Yes. Can you improve on the component problems? Yes. Do we collectively have the ability to make the world more just and sustainable for humans and for wild species? Yes. Can you learn to cope, survive, and even thrive in this era? Yes! That is the radical thesis of this book.

I will show you how to reframe eco-anxiety so it works for you, not against you. I personally return to these methods again and again even as I coach others around the world to do the same.

THE PROCESS OF THE BOOK

In this book, I will share with you the program that I use for both clients and therapists alike to orient themselves in the truth of our current climate crisis, to be able to manage their own reactions to it, and plan for a future of positive personal growth. In expressing your feelings, you can actually shape them, and make goals for how you *want* to feel. I'm going to teach you how to do this and, in so doing, liberate your ability to thrive and feel happiness—in nature as it is, even damaged; in your relationships, however fraught; and within your community, which is likely polarized. The therapeutic program I practice and teach is designed to find positive solutions. It really is possible to feel legitimately happy. It is possible to feel reconnected to the people and places you love. It is possible to feel energized and excited about change—in yourself and in the world.

It has been said that "the purpose of psychology is to give us a completely different idea of the things we know best."[5] And in this book I will focus on the fundamentals of your mental health: how you think, how you feel, how you manage stress, how you create your identity and sense of self, how you dream, how you communicate and create healthy relationships, how you take action on your values, how you can change and grow as a person—and how you can ultimately let go, accept your mortality, and see your wild and precious life in a transcendent way. What's unique is that I talk about these things in the context of some of the defining environmental issues of our time.

In Part One, to tackle global-scale issues we start small, with basic coping skills: how to think, how to feel, how to manage stress, and how to take a clear-eyed look at how climate and environmental issues actually affect you, by imagining your own personal IPCC—"Individual Problems with Climate Change"—report.

Then in Part Two I help you think about your identity and the core values you hold about nature, and how this manifests in your family and culture. This creates a solid and healthy base for coping.

From there, in Part Three I will teach you skills drawn from therapy to deal with troubling issues that threaten your coping, like eco-anxiety and despair and grief about the destruction of the natural world. This includes how to handle politics and propaganda that inflame eco-anxiety, and how to acknowledge the land where you live, whether it be healthy or degraded.

In Part Four I'll help you reclaim happiness in an ethical way that fits with your values. We will envision what well-being and flourishing would look like for you. This might be in the realms of nature and the outdoors. It includes your relationships and "eco-friends." You'll learn ways to find solace and inspiration in creativity and the arts. We'll step back and consider the importance of religion and spirituality in your coping, identity, and thriving. This includes ways to transcend your own small self and experience your being within a larger sweep of nature and generations.

Then in Part Five, we move into action—your unique style, not anyone else's; the options you have; how to troubleshoot when things don't work; and what to do when disaster arrives at your door. What comes next? Being a "master of two worlds"—being better able to hold and balance the world you strive for with the world you have been given.

The process I outline in this book is in reality a coping cycle, a program that you can use now and return to again and again in your life. It is also a set of insights and tools to share with your family, group, organization, or community. In the coming years, you'll find that when you're in the midst of challenges or have embarked on a project or important life goal, you can remind yourself to be conscious of how you think, how you feel, how you manage the inevitable stress, and of your identity and core values. That brings you back to the beginning of the coping cycle.

HOW TO USE THIS BOOK

You can use this book yourself as a private growth experience. In many ways the book mimics the arc of a counseling or therapy experience: listening, caring for you, believing in you, and giving you the opportunity for self-exploration. The reflection you'll do and skills you learn will build on themselves and take on larger and more meaning later.

You can also work through the book with a group, and I encourage you to share the process with others. There is a universality to the topics in these chapters. They're doorways to conversation, mutual validation, and the diverse ways environmental identity manifests.

After learning the tools I offer, many people find that the hard things—the eco-anxiety, the environmental grief, and the feeling of "why bother?"—shrink to what they should be: healthy and normal emotions that guide us and that are entirely appropriate for our times. This may be your experience. Some people need more help: Their distress, despair, or trauma requires major efforts at healing, or they are enduring a crisis that makes thriving and growth impossible at the moment. That's why ecotherapy and disaster mental health approaches are also covered in this book.

ABOUT ME

I'm what is known as a climate psychologist (technically, I'm a clinical psychologist and mental health therapist with a specialization in environmental psychology, a rare combination). The connection between nature and mental health, and unavoidably the problem of climate breakdown, has been the focus of my work for decades.

People often ask, "How did you find your way into this work?" I sometimes joke it's the other way around: Climate change found me. It wasn't my goal, but in some ways I have been cross-training myself for this precise moment my whole adult life. I came to clinical psychology after early experiences as a professional fisherman in Alaska, a river guide in the Grand Canyon, a Greenpeace advocate, and a counselor on a therapeutic wagon train traveling the American West with inner-city teens. As a grad student, I studied health psychology while treating anxiety and depression in cardiac rehabilitation patients, and I trained in mindfulness meditation and behavioral medicine. All of this comes together in my approach to ecological issues in psychology today.

I'm not a passive bystander. I am a parent. I grew up in a place prone to extreme winters and now live where drought and wildfire are yearly threats. The journey of climate coping is my journey too. I have had to grapple with nights made sleepless by catastrophic visions, and sit with frustration and anger while we deplete our fresh water and exhaust the bounty of our forests and oceans at an exponential rate. I have learned that loving the Earth will break your heart. I'm still adjusting to the loss of the Holocene era (and taking for granted that climate, places, and seasons will always be there). I'm also a widower. My wife was a youthful victim of metastatic breast cancer, diagnosed at age thirty-five, in otherwise vibrant health, and her untimely death raises questions about the ecological risks we face. I completed this book as a single parent helping my teenage daughter through her high school routine. I know what it's like to grieve while also worrying about my child's well-being.

I have a specific viewpoint and lived experience as a white, educated psychologist from North America, born in the late twentieth century and writing in the early twenty-first century. I'm shaped by my culture and privileges but grappling with questions I believe are universal for many readers in many places and times.

In other words, no matter who you are, we have much in common. We are all walking this new climate change path together. It is the road all of us privileged with being alive right now must take.

I draw on some approaches in health care that aren't typically applied in relation to eco and climate impacts. For example, I am guided by the *harm reduction approach* used in public health and addiction treatment: Even if the problem cannot be removed, the goal is to reduce the health damage as much as possible.[6] In the short term, we cannot change the basic realities of the eco and climate threats we face. But we can get better at being more resilient to them and limiting their damage.

I am influenced by the concept of *palliative care*, an approach to healthcare aimed at optimizing people's quality of life and alleviating or reducing suffering for those with serious, complex, or terminal illnesses. From a palliative care perspective, aid and comfort can be provided at any point in the process of injury and illness. There is no giving up and no need to, even if there's no solution or "cure" for the ailment.

Drawing from science advocacy, I adopt the stance of an *honest broker*.[7] I am not going to tell you what to think, or feel, or do. The readers of this book are diverse, and the eco and climate situation will keep changing. I will, however, share many options for how you might think, how you might develop your emotional intelligence, and how you can find a path for action that is fitting and authentic for you now and can be adapted as you learn and grow.

My focus is on *post-doom*, what to do *after* you have your ecological wake-up—when you realize you're still very much alive and have some time and opportunity to make a difference in your own life, for others, or for what eco-philosopher David Abram calls the "more-than-human-world."[8] You won't find preachy, gloomy, data-heavy, desperate, and finger-pointing "us and them" language in this book. The feelings we are going for are compassion, forgiveness, love, and creativity. The mindset is an invitation to growth, mindfulness, humor (as appropriate), acceptance, commitment, and self-transcendence.

THE CHARACTERS

Through the various chapters, I'll introduce you to people who want to feel better but don't know how to begin, facing tough questions about how to protect their family, where to live, and how to find integrity in their careers and lives. If you see yourself in these characters, that is not surprising. While they are composites, they are populated with living details from real people I have worked with.

You'll meet a nursing mother up all night doomscrolling on her phone, worrying about how to protect the child she brought into the world; an urban city planner worried about his blood pressure as he works to protect the African American community; a teen activist feeling isolated from peers who are tired of listening to her sounding the alarm; a glaciologist struggling to cope back in suburbia after witnessing the collapsing ice sheets; an elder environmentalist depressed about what he fears is a legacy of failure; a mental health counselor who feels ill-equipped to help clients in despair over the environmental crisis; a wildlife veterinarian studying "forever chemicals" and coping with summer depression; a political organizer who hides her discomfort with going out into nature; a promising twentysomething who wonders what it was like when people believed in a future; partners estranged over how to show their nature values and deal with issues like plastic products and recycling; a young woman struggling to grieve the loss of her family's cherished woodland to catastrophic wildfire. You and I will walk the path with them to learn from their struggles, witness their growth, and gather insights for your own life.

As I worked on these pages, I found solace and inspiration in the stories of others who have wrestled similarly in the past—for example, Henry David Thoreau, who wrote in the 1840s, another time of rapid technological change, environmental concerns, and struggles for justice. For Thoreau, the disruptive new technology was a commercial telegraph that suddenly created a hunger for the latest national and international news. "What news!" he observed. "How much more important to know what that is which was never old!"[9]

This book is meant for readers who have a pressing need for a path through eco-anxiety to environmental identity and their proper rights and duties in relation to themselves and the Earth. If the chapter titles sound timeless—"Thinking," "Feeling," "Values," "Nature," "Family," "Despair," "Happiness," "Art," "Spirit," "Duty"—it's because they are. This work is meant to be sustainable for you, for the long haul. Wherever you are, please accept such portions as apply to you. Echoing Thoreau, I trust that none will stretch the seams in putting on the coat, for it may do good service to those whom it fits.

So it is with respect to Earth, to the many people and ideas I draw from, and to your one and precious life that we begin.

Surviving Climate Anxiety

PART ONE

Coping

MANY PEOPLE ASSUME THEIR OPTIONS AND ABILITIES TO take on environmental problems are limited. Given the scope of the issues, happiness and thriving can seem out of reach. But when you focus on curiosity and creativity, let go of the need for mastery or perfection, and develop a willingness to try new things, you can create your own version of coping and thriving in an era of climate change. Coping is not just about feeling better. Having effective emotional coping skills makes it more likely you'll learn the lessons you need to address the problems you face.

The theme of Part One is "Open mind, open heart, open breath, open hands." The chapters in this part mirror the process I use in my first meetings with people experiencing anxiety and despair related to climate issues and other environmental stresses. In these chapters, I'll show you how to approach these difficulties with a *growth mindset*. I'll help you expand your emotional capacity to bear the situation and increase your *psychological flexibility* so that you can respond to stress without being overwhelmed. This will prepare you to take a brave, honest look at how climate and environmental threats actually impact you, so you can plan and act accordingly. Once you get some breathing room (literally and symbolically), I'll help you look at your environmental identity and values in a deeper way.

Chapter 1

Thinking

---

*"Letting something in" is too passive; what I'm talking about is
fitting a hyperobject into your heart without it breaking.*

—DANIEL SHERRELL[1]

THE GLOBAL CRISES THAT FILL THE HEADLINES ARE so complex and massive that they can completely overwhelm you. The threats are insidious, seeming to come at you from all directions, along with disagreement about the problems and solutions. Old language becomes insufficient. We need new categories like "unnatural disasters" and new feelings words like "solastalgia" (the sense of loss about disappearing nature).[2]

Dealing with complex problems like climate change reminds me of the parable of the blind people and the elephant, but on a global scale. In the story, one blind person touches the trunk of the elephant and declares that the animal is a curving tree branch. Then another blind person pats the elephant's side and says it's like a wall. The blind person by the leg perceives the elephant to be a tall pillar; another pulls on the tail and declares that the animal is like a rope. Our understanding of the elephant of climate change is similar: One person sees a scientific problem to solve, another a matter of economics; another sees injustice and political dysfunction. Another sees growth opportunities in green energy, while another sees the need for a new Earth-based spirituality. Amid the crisis, life seems to go on as before.

Understanding that we can only perceive our part of the bigger picture when it comes to climate change helps us recognize that we each feel and think differently about the threats we face. Some people are being crushed under the elephant's feet, while others are seated comfortably on the elephant's back wondering what all the catastrophizing is about. If you've picked up this book, you're likely experiencing anxiety about climate change. I see you, and I see this in my psychology practice every day. This anxiety manifests in many ways, none of them "right" or "wrong."

Throughout this book, I'm going to show you new ways of thinking about how global climate change and other environmental issues are impacting you emotionally. I want to be clear that I'm not going to tell you *what* you should think or what you should feel. But I am going to teach you *how* to manage your thoughts and feelings, so that you have the tools to cope with and ultimately overcome the unhealthy aspects of climate anxiety.

There is a catchphrase I use in climate therapy: "Validate, elevate, create."

- *Validate.* No one wants to be seen as weak or too emotional or alarmist. To work through your fears, you need the space to acknowledge that they are valid. You should not be ashamed of your sense of alarm and vulnerability.
- *Elevate.* I want you to know that your climate concerns are not a peripheral source of anxiety. They are central, vital, and absolutely worthy of our attention. In this book, we'll practice putting them first.
- *Create.* Once your senses are no longer narrowed from threat, you will be able to activate your creative energies and curiosity to open your mind to possibilities for coping.

THE WAKING-UP SYNDROME

A sudden awareness about just how bad environmental problems are can be so completely disorienting, therapists even have a term for it: "waking-up syndrome." I know how this feels. I've been knocked off balance by environmental wake-up calls many times over the years, like during my time working for Greenpeace, or my first foray into climate disaster research. In these moments, you realize the massive scope of the issues and feel a crushing sense of responsibility to make a difference. You struggle over the smallest daily decisions. And you begin to wrestle with existential questions. Should you see as much of the world as you can before it's gone? What kind of planet are you leaving for your children to inherit?

Larysa

Larysa was a thirty-year-old mother on maternity leave from her job as a project manager at a large tech firm near Seattle, where she was known for her ability to define the scope of a project and marshal the resources to get it done. It was the COVID-19 era. In our first online session, the dark smudges under her eyes betrayed her fatigue. She had thick dark hair that was pulled neatly back, and a square face with high cheekbones that hinted at her eastern European heritage. At her back was a whiteboard that her husband, Declan, an Irish programmer originally from the suburbs of Dublin, had set up for his own work Zoom meetings.

Larysa had contacted me for therapy because she felt increasingly paralyzed by anxiety about the climate-change-related storms and disasters where she lived in the Pacific Northwest. She told me she was frequently up all night, nursing her new baby while staring into the bright light of her phone screen, scrolling through a phantasmagoria of disaster stories, each more frightening than the last. She was so worried about losing power during severe winter storms that when Declan asked her what she wanted for her birthday, she requested an emergency generator.

The get-it-done mentality that made Larysa so good at her job led her to approach her climate disaster concerns as problems to be researched and solved. But now she faced a hard-to-grasp and ever-widening problem that she couldn't project-manage her way through.

When Larysa thought about the threats of climate change, she felt she had to take action. Ignoring the data felt like denial. All she felt like she could do was vigilantly keep up with the news. Otherwise, she told me, she felt like an "ostrich with her head in the sand."

I've worked with many people like Larysa, confident in many aspects of their life but at a loss for how to deal with complex, existential issues like climate change. After listening to her story, I immediately knew that there were several things that needed to happen before she could begin to use her skills to address the issues that were troubling her. My validate-elevate-create approach suggested a path forward:

1. Larysa was isolated and needed validation by someone who took her eco-concerns as seriously as she did. Further, she needed to understand why certain issues affected her so powerfully.
2. She needed to accommodate and better understand the magnitude of systemic problems like global climate change. And she needed emotional skills to navigate them.
3. She had to find points of leverage that could make a difference: small, doable actions that would have a measurable positive effect on her mood and self-esteem and on the well-being of her family. But first she needed to get past a limiting mindset about what she could accomplish.

Larysa and I began by honoring the ostrich.

It's actually a myth that ostriches hide their heads in the sand at the approach of danger. They either run—what do you think those long legs are for?—or camouflage themselves. And when danger arrives, an adult ostrich doesn't hide; it strikes back with a clawed foot that delivers a kick powerful enough to kill a lion. The myth is born out of the fact that ostriches build their nests and lay their eggs in the sand, so when an ostrich mother has her head in the sand, she's likely turning over her eggs, tending to her young. I gently suggested that Larysa, as a postpartum mom, was herself behaving like an ostrich: tending her young and working to keep them safe.

Reframing Larysa's image of the ostrich was enormously helpful to her. I'm a parent too, and I share Larysa's worries about climate change and the dangers posed by storms, heat, and wildfires. Researching disaster scenarios is a real parenting responsibility—a kind of tending that is necessary in the twenty-first century, and one that may allow us to respond more quickly if those scenarios come to pass. But for Larysa it became an unhealthy coping skill. Her natural obsessiveness, along with her heightened postpartum worries about how to protect her child, led her down a rabbit hole of increasingly extreme disaster preparedness ("prepper" websites). Ironically, Larysa's screen habit made it harder to be present for her child. Each doomscrolling session left her feeling more unsafe and inadequate as a mom.

YOUR LIFE IS THE NEWS

When it came to getting creative about Larysa's situation, the need for healthy coping skills was the first step, starting with setting boundaries about her disaster-prep research and her screen time. An immediate practice that Larysa implemented was leaving her phone in another room at night when she was nursing. This gave her some time every day when she was not being agitated by news and could be in the present moment for herself and her baby.

My standard recommendation is to take daily screen breaks to clear your head, particularly from addictive phones and devices, and I practice this myself. Taking a break from her phone habit was one way for Larysa to find space to think more clearly and find immediate relief for the anxiety she was feeling. I reassured her that in an actual emergency she could access an unprecedented amount of information within a few clicks. The key was that she, not marketing and social media algorithms, was in control of her search for information.

The next step was putting the sense of impending disaster into perspective in the context of the lived experience of Larysa's daily life. The global scale of issues can be dizzying, leading to a sense of *scale vertigo*.[3] So, an important skill is learning to look away from the headlines and focus on what is happening in the here and now. "Larysa, *your life* is the news," I gently informed her. I invited her to close her eyes and take a breath. "Imagine you are reporting on your own life. Step outside your door. Use your senses. Look around. What do you see?" Her list included a modest, well-built house in a quiet neighborhood, a small garden, a healthy child, a loving spouse, friendly neighbors, supportive extended family, and an interesting job.

Yes, the house needed paint and the garden she had tended during her pregnancy had become overgrown. But when Larysa compared her climate fears with a real-time perspective, she was able to recover a feeling of *present-moment safety*. On that day, in that moment, she was not in immediate danger. In fact, she could recognize that a more immediate danger was her late-night doomscrolling, which affected her health and took her attention away from her child.

One of the most important turning points in my early meetings with Larysa was a comment I made in passing that for her children "there will be good days." We can be sure that there are bad days to come in terms of climate disasters. That's realism. But I do believe there will also be good days, when the sky is clear, the weather is gentle, and you can enjoy your life. This also is realistic. I hope that by the end of this book you will feel confident to give yourself permission to enjoy the good days you have, find meaning in your life, and take action to make good days a possibility for as many people we can.

When Larysa thought of the elephant parable, she felt relief. It was impossible for her to know what's going on every second of the day on every front, impossible to see all parts of the climate crisis at the same time. And that was okay. Accepting that her vision was limited to what she needed to focus on as a parent was liberating. Not only is it impossible to know what's going on every second of the day on every front, but in Larysa's case her obsession with staying informed was harming her mental health. Letting go of her impulse to see—and, underneath, to control—the whole elephant was a kind of release. It set the stage for a fresh start.

I suggested that the next time Larysa was up late at night nursing or when her baby, Sammie, couldn't sleep, she could take Sammie outside and look up at the sky. The storms and fires of climate change were still there in the world, and she would work to address them tomorrow, and in the months and years to come. But for now, she could enjoy the simple pleasure of sitting outdoors at night, wrapped in a blanket with her child, like many mothers before, singing soft lullabies and gazing at the stars.

Roman

I met Roman at a climate science conference. He was a compact and fit young man in a fleece vest, with quick, precise movements and intense green eyes. As a doctoral student studying glaciology, Roman had just returned from his first expedition to Antarctica, where he had studied changes in the rapidly melting Thwaites Glacier—nicknamed the "Doomsday Glacier" because its dissolution will drive catastrophic sea level rise. He asked about my name tag and wondered what a psychologist was doing at the meeting.

We met for coffee after Roman's presentation. I asked him about his recent experience of being at the bottom of the world. I knew on an abstract level that melting polar ice caps cause sea levels to rise. But Roman allowed me to see the problem through his eyes. He asked for my laptop and navigated to a map view of Antarctica, showing Earth from the perspective of the South Pole—a view of the continent I had never actually seen. He pointed

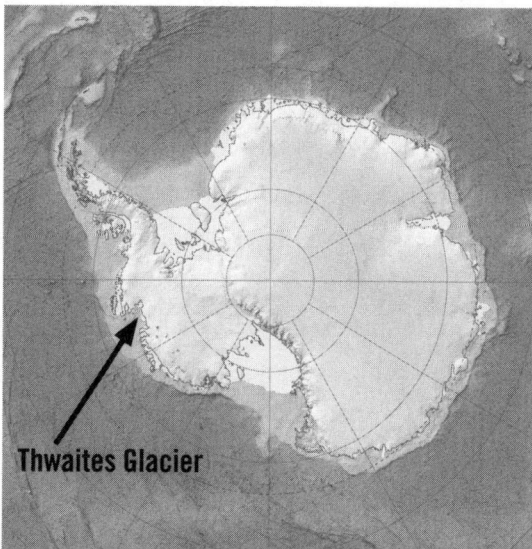

Thwaites Glacier

out the large, relatively stable East Antarctic Ice Sheet and the smaller, less stable West Antarctic Ice Sheet, separated by the Transantarctic Mountains.

Then Roman pointed to the Thwaites, the widest glacier in the world at 74,000 square miles—roughly the size of Great Britain or the state of Florida. When viewed from a plane, it stretches as far as you can see. But the glacier is highly vulnerable as the ocean warms. As Roman explained, much of Antarctica lies under sea level, pushed down by the miles-deep ice. As the floating ice shelves protecting glaciers wear away, warmer ocean water seeps under the glaciers more quickly, undermining them from below. It is a collapse of the Thwaites and even the whole West Antarctic Ice Sheet that scientists are concerned about. The melting of the Thwaites Glacier currently contributes about 4 percent to sea level rise. If it were to collapse entirely, it would set up a chain of events that would add two feet to sea level. Roman told me that according to some data, the Thwaites could collapse within a decade. I took a sip of cappuccino and held his gaze, letting the implications hang in the air.

Upon his return home from Antarctica, Roman was haunted by what he had seen and learned. He was despondent and felt isolated when he realized that most people didn't fully appreciate what the loss of the glaciers and ice sheets meant for the world. With so many other issues demanding people's attention, how could he help them see what was going to happen in Antarctica? Even his fiancée, Sandra, whom he had met in college in a human geography class, and his friends, themselves scientists and academics, had a hard time making these faraway events a priority when they were focusing on other pressing issues such as wars in the Middle East and racial justice at home.

Roman had a severe case of waking-up syndrome, the sense of confusion that follows a painful epiphany or consciousness-raising about the state of nature. Now that he was awake to the potential for cataclysmic disaster, he didn't know how it was possible to enjoy even the simplest things in life. Like many people I have met who have returned to urban society after spending time deep in nature, Roman struggled with the sounds of traffic and leaf blowers. Aside from being noisy, they only reinforced his new vantage point, from which he saw all of society as one huge threatening machine.

He began to scrutinize every decision he and Sandra made, and wrestled with the question of whether to continue with his doctoral studies. Was it ethical for him to add to greenhouse gases by taking plane flights around the globe? He also saw how months-long absences on "the Ice" took a toll on his colleagues' relationships and families and wondered if it was something his relationship could withstand. He also worried that his social drinking was becoming a daily habit, especially given that there was a history of alcoholism in his family. He knew this wasn't a healthy way to cope with his anxiety. Should he really be sacrificing his relationship, his health, and years of his life on this work? After all, scientists, including his own professors, had been raising concerns about the melting ice caps for years, and where had that gotten us? Did adding to the science even matter? If not, what would he do with his life?

HYPEROBJECTS AND WICKED PROBLEMS

How can we enlarge our thinking to take in topics as big as Earth and as long as history? The ability to zoom out to see the globe from a million feet (as if we were on the International Space Station) while simultaneously zooming in on our daily lives requires not just thinking but also *thinking about our thinking*. In other words, when dealing with meta-challenges, we need meta-cognition. What are the right thinking tools for the job?

Upon his return from Antarctica, Roman, isolated in his suburban neighborhood, was not just tortured daily by the knowledge that those noisy leaf blowers were powered by fossil fuels. He saw all the connections between the leaf blowers, the plight of the low-paid immigrant workers who were wielding them, the global economy, global heating, the collapsing ice sheets, and catastrophic sea level rise. This was pure torture.

In his private existential crisis, Roman was confronting the *hyperobject* of climate change. "Hyperobject" is a twenty-first-century term coined by philosopher Timothy Morton to describe a thing so enormous that it defies time and space.[4] Global climate change, capitalism, nanoplastics in the environment and our bloodstreams, the proposal to establish a new geological age called the Anthropocene (the era when collective human activity has become the dominant influence on Earth's climate and the environment)— all these things are impossible to perceive in their entirety. It's like the climate elephant. No matter how hard we try, we can't fully get our heads around these concepts. Even if we could, it would be impossible to take an objective view because we are part of them.

People who have the opportunity to orbit Earth and look upon our planet from space commonly experience the "overview effect," a breathtaking sense of awe and reverence when seeing Earth as a whole. But glimpsing the whole can also be a shock to our mental health. The shadow side of planetary awareness is an ominous *dark ecology*, another term coined by Timothy Morton for hyperobjects and the terrible truths they hold—in this case, the dark truth of planetary changes making swaths of Earth inhospitable for humans and other species is made all the darker by the knowledge that these changes are being driven by our own behavior. Being engulfed in dark ecology brings on a sense of guilt and hopelessness.

When I first began studying environmental issues, I came across a useful term: *wicked problems*, used for big, complex issues that defy simple cause-and-effect solutions, like poverty, violence, and disease. Wicked problems are actually a bunch of issues tangled together. Climate change and environmental crises are wicked problems not only because they are hard to navigate but also because when we tackle one part of the problem, we may make a related issue worse. (Looking ahead, I'll help you think of *wicked solutions* for these problems.)

CAPITAL-*I* VS. LITTLE-*I* ISSUES

Some global issues command your attention more than others. When you have waking-up moments, you don't always know the full scope of what you are confronting. So you begin with what you know; these are the things that stand out, or (in psychology terms) have more *salience*. A new mom like Larysa focuses on keeping her children and family safe. Scientists like Roman focus on their specialties: in his case, glaciers. Psychologists like me focus on the emotional fallout of the climate crisis and how to help people grapple with their anxiety and live according to their values.

A deeper, more insightful form of coping begins with the realization that we all have capital-*I* issues as well as little-*i* issues. What I call *capital-I* issues are the big, global-level problems we focus on, captured in slogans like "End Hunger" and "Save the Planet." *Little-i* issues are our personal-level stuff, our private traumas, neuroses, and whatever psychic baggage we carry. Little-*i* issues also include how the big capital-*I* problems show up in your personal lives: guilt you feel about your carbon footprint, the insomnia you suffer from as you ponder the future of the planet that is being left for our children. The capital-*I* issue for Roman was the Doomsday Glacier and everything that he knew as a scientist. His little-*i* issues included his relationship with Sandra and his stress-related drinking. Traditionally, a therapist's job is to focus solely on little-*i* issues. In climate-conscious therapy, I help people tackle both the capital-*I* and the little-*i* issues at the same time. *The personal is planetary* when it comes to protecting your mental health in the era of climate change.

Agnes

Agnes was seventeen when she came to see me. She had tousled sandy-colored hair and wore baggy denim jeans, scuffed Chuck Taylors, and a weathered Bob Dylan T-shirt she'd found at a thrift store. Already she was a seasoned environmentalist, participating in climate walks and protests against the expansion of the local freeway. She had been speaking to the media about climate issues since middle school.

Naturally skeptical of people over thirty, at our first session Agnes openly wondered if I was like other adults who were unable to grasp the high stakes of the "global polycrisis" (she was demonstrating her knowledge by using a newer technical term for many wicked problems occurring at the same time on a global scale). Was I like her previous therapist, who didn't know what the Keeling Curve was (the trend of rising atmospheric carbon dioxide measured at the Mauna Loa Observatory since 1958)? Demonstrating that I knew the terms and that I could see the world as she saw it established a foundation from which I could help her deal with the feelings underneath her brave facade.

New stories of environmental injustice plunged her into depressed moods. She tried to stay busy to avoid the worst feelings. She felt left out when she couldn't take a break from her climate mission and live in the moment with her friends and have fun. Recently she had begun to feel hopeless and burned out. To cope, some nights she binge-watched streaming shows while polishing off pints of ice cream. This made her feel even more guilty.

Young people, including Agnes, have told me of being scared and guilt-tripped from their earliest years at school, even feeling responsible for the disposable diapers they wore as babies. In addition to feeling fear and guilt, psychotherapists like Caroline Hickman have documented that many young people around the world feel betrayed by adults for not adequately address-ing the climate crisis. This sense of betrayal is primarily directed at people in power who have stifled global action, but it shows up at home too. Agnes wondered how her parents could not see that if the climate crisis is a killer of humans and other animals, then their daily carbon emissions are like fingerprints on a murder weapon.

For all her climate change knowledge and precocious empowerment, Agnes was still a teen. Like many adolescents, her thinking could be con-crete, idealistic, rigid, and judgmental. She was high-achieving and at times self-punishing. When I began to talk about capital-*I* and little-*i* issues with her, it helped her acknowledge that despite her outward confidence, she often felt like a privileged imposter. Despite her growing knowledge, she was still discovering the world.

BRIGHT ECOLOGY

I sat with Agnes, reflecting on how a central aspect of the waking-up syn-drome for young people is their first harsh awareness of dark ecology and sense of polycrisis. When this happens, their sense of interconnection with nature tragically morphs into fear of an insidious threat. And when young people realize their daily acts are embedded in an unsustainable, seemingly planet-destroying system, opportunities for guilt and shame are omnipres-ent. I worried for teens like my daughter, who weren't obviously suffering from such disillusionment as Agnes was, but still have to learn to live in a flawed and unjust system.

Now that the feelings were present in the room, how could I help Agnes be creative in her thinking? My own political and ecological awakening occurred when I was much older than she was. What would it have felt like to confront the end of the world when I was her age?

First, I validated. I told Agnes it was only natural that she would feel a mix of surprise, astonishment, affront, anger, disappointment, betrayal, and probably responsibility, guilt, and some blame as well. "And I'm sorry to say it's worse than you think," I told her. "You're going to find even more wicked problems the deeper you go."

Then I said, "So, here's the creative piece. It's also better than you think. From where I sit, there is also innovation and hope for you, Agnes, the kind that didn't exist when I was a kid in the 1970s. Yes, my Gen X crew grew up with Earth Day, awake to humans' poor stewardship of Earth. Growing up, my generation still didn't understand the scope of what we faced, or the stakes. But we also didn't understand the possibilities."

I explained to Agnes that the flipside of dark ecology is *bright ecology*, a companion term I've coined for the amazing opportunities for growth and hope that our challenges present. I have seen with my own eyes that there are legions of individuals making the most of these opportunities in their work and lives through invention, government policy, grassroots efforts, and the arts. And Agnes agreed. The unexpected bonus of being informed and engaged with activism is that she meets creative, inspiring people around the world who share her values and, together, they celebrate small and large victories. Once you wrap your mind around the fact that things are better than you think, you can experience genuinely positive feelings—humor, joy, inspiration—along with the anger and guilt. Agnes had found a climate-conscious community that understands her and vali-dates her concerns. It's not *the* solution. But it is part of the solution. Finding and engaging with this community is vital for coping.

A GROWTH MINDSET

How you *think* about climate change profoundly affects how you *feel* and *act* about climate change. A *growth mindset* assumes that your skills and abilities are not fixed, and there is always more you can learn with time and practice.[5] It's okay to admit your own lack of knowledge and the limitations of what you understand. In contrast, a *fixed mindset* about eco and climate issues makes assumptions that limit your potential and ability to change and learn new things: "Climate change is too big a problem to solve. It's too late to do anything. My actions don't matter." It's easy to fall into beliefs like this, called an *anchoring bias* in psychology because they anchor you to a certain set of possibilities. The strongest unique predictor of intention to take climate action is whether a person believes that other people are already taking that action.[6] However, you might look around and assume that avoiding talking about the crisis of climate change is the norm. Indeed, as the Yale Program on Climate Change Communication reports, we (and especially those of us in the United States) tend to underestimate the percentage of global citizens who want to see action taken on climate change.

Cultivating a growth mindset is essential to your climate coping because it invites flexibility of thought. It creates conditions for awareness, mindfulness, openness to new ideas, and the will to take action regardless of a guaranteed outcome. A growth mindset will therefore be your ally as you open up to eco-distress. The key insight for Larysa, Roman, and Agnes, and for you: *You don't have to solve the climate crisis in order to cope with it.* If you have a growth mindset, the global environmental crisis can be thought of as a series of challenges that will be met over time, in your life and beyond. Your job is just to show up each day and do a good enough job being the best person you can be.

The novelist E. L. Doctorow has said: "Writing is like driving at night in the fog. You can only see as far as your headlights, but you can make the whole trip that way."[7] The same is true of our journeys through climate change: You don't need to be able to see far into the distance as long as you can see what is directly ahead. Think of Larysa's, Roman's, and Agnes's climate awakenings and discoveries not as a wall but as a doorway—even a *rite of passage*. In a rite of passage, not everything can be known, but *everything can be felt*, and that brings us to Chapter 2.

Chapter 2

Feeling

I was…one of those most afraid, and by pretending to be brave,
so as to encourage the others, I grew brave.

—Lotte, in *Sorrows of Young Werther*,
by Johann Wolfgang von Goethe

Succumbing to climate anxiety and despair is not inevitable. The challenge is learning to see your feelings about eco and climate issues as natural experiences to trust and learn from. The process begins with emotional awareness and expression and moves to *emotional regulation* (think of a pilot wrestling with the controls of a spiraling airplane to bring it back into level flight). And emotional expression and regulation are the basis for *emotional intelligence* about your responses to eco and climate issues. Emotional intelligence, along with a growth mindset, and the stress management skills covered in the next chapter, form the backbone of your ongoing adaptation response to the climate crisis.

To begin, I want to ask you three big questions about emotions, the same questions I pose to clients. The first is "What exactly do you feel about climate change and the other environmental issues that affect you?" The second is more radical: "What feelings do you *want to have* about these issues?" And the third question is more profound: "What feelings do you *need to feel*?" When you are stuck in a rut, it can be a very helpful ability to redirect your feelings. Yet sometimes you need to resist that urge and stay with the discomfort. That's where the wisdom lies.

Delphina

Delphina grew up in a large Greek family in the suburbs of Chicago, near the shores of Lake Michigan. She was a high achiever pursuing a doctorate in veterinary medicine, and her family was proud of her accomplishments. Most of us think of veterinary medicine as the business of treating household pets, but today's veterinarians study a range of important ecological issues, including food safety, animal-borne diseases like COVID-19, and wildlife protection, and Delphina was dedicated to all of them. In particular, her research led her to confront the health dangers of per- and polyfluoroalkyl substances (PFAS), "forever chemicals" that accumulate in the tissues of fish like the wild walleye and yellow perch caught in the Great Lakes, and how this affected the local food chain and ultimately human health.

Delphina came to vet science out of love for aquatic animals like her namesake, the dolphin. Now she was experiencing waking-up syndrome due to her newfound awareness of the toxic legacy of the industries that her family had worked in around the Great Lakes—one that now led to warnings against eating wild fish, frogs, crayfish, and even deer from in or around contaminated areas. The more Delphina progressed in her studies, the deeper her environmental despair became.

It was mid-June when we first spoke. As a child, Delphina had always gleefully anticipated the seemingly endless carefree days of summer after the snowy Midwest winters. But she no longer trusted summertime. She associated it with a dark ecology of stifling heat, warming lakes, harmful algae blooms, oxygen-starved dead zones from agricultural runoff, and recreational boats that transported exotic invasive species. These days, she felt tortured by the heat of the long days and exhausted by the bright sun, had trouble sleeping, and had begun to feel depressed. Delphina was also dealing with some inner struggles. She was feeling increasingly certain that she did not want to have children, now and maybe ever, and had been discussing this with her partner, Gabriela. Delphina had already gone through coming out as queer to her very traditional "big fat Greek family," who to their credit were surprisingly supportive of her. But the presence of generations of children, grandchildren, nieces, and nephews was so ingrained in her family, she dreaded sharing her reluctance to have children with them.

I got to know Delphina along with several other students from around the world who served on the committee on wellness for the International Veterinary Students Association. All the vet students were witnessing the declining health of species around the world. Bidisha lived in Mumbai, India, where she was helping to prevent indigenous cattle from going extinct. Kamau was studying zoonotic (animal-borne) diseases that affect poor rural people who lack access to health care and sanitation in his home country, Nigeria. Much like Roman, the glacier researcher, they felt isolated and struggled to convey the importance of what they were seeing and learning to their friends and families. As I listened to their stories in our online meeting, my eyes landed on a framed postcard on my office wall, with a quote about validation from writer Kurt Vonnegut: "Still and all, why bother?" Here's my answer: I feel, think, and care as much as you do about many of the things you care about, although most people do not care about them. You are not alone.

Big issues like climate change left Delphina stuck in an emotional rut. She couldn't enjoy pleasant moments with friends and family or while out in nature on a beautiful day. She was feeling *anhedonia*, or lack of pleasure in typically pleasurable things: a common but underrecognized side effect of environmental despair. Anhedonia is also associated with clinical depression, and addressing her anhedonia was critical to help prevent Delphina from lapsing into a depressive disorder. (I'll talk about coping with eco-depression later in the book.)

As with Larysa, the new mother staying up all night nursing and doom-scrolling, I had immediate recommendations to help Delphina feel less desperate in the short term. To improve her overall mood, I recommended better sleep hygiene, including going to bed near sundown instead of staying up late under bright lights and staring at bright screens. I also suggested blackout drapes for better morning sleep and reducing her late afternoon iced coffee and evening wine cooler consumption, caffeine and alcohol being known sleep disrupters. To help her reconnect with some of her care-free memories of summer, I encouraged her to take moments to pay attention to simple things like the sunset, listening to birdsong, and watching clouds. I also recommended she schedule a week when she could take an old-school summer vacation, away from the news and computer screens, and take walks, have bonfires, write in her journal, and read poetry or novels—both recently lost pleasures. Then we got to work on her troubled feelings.

WE HURT WHERE WE CARE

When what you love or value is under threat, you feel it. "We hurt where we care," as the saying goes in therapy.[1] Your emotions—and the words and body language you use to express them—are the primary, immediate way you experience upsetting climate and environmental issues. In fact, strong feelings about climate and environmental issues are quite common. More than half of Americans report they feel curious, fearful, anxious, or angry about climate change, while less than a third feel optimistic. While young people are believed to experience stronger emotions about eco and climate issues than older people, the stereotype of an age gap is inaccurate. People of all ages report similar levels of concern.

From my perspective as a clinical and environmental psychologist, there are three basic psychological tasks associated with addressing environmental issues and other social problems. We must undertake these individually, as communities and as societies:

- *The expressive task.* Recognizing and sharing our feelings about the issues or situations, and developing emotional intelligence about our and others' emotions.
- *The descriptive task.* Developing an intellectual understanding of the problem(s) through research, personal experience, data, and images.
- *The prescriptive task.* Deciding what to do, promoting strategies and policies for change, and following them through until the crisis is overcome.

Expressive Task
Recognizing and sharing
feelings about an issue
or situation, developing
emotional intelligence
about your and others'
responses

Descriptive Task
Developing a comprehensive
understanding through
research, personal
experience, data,
and images

Prescriptive Task
Deciding what to do,
promoting strategies and
policies for change, and
following them through

Addressing Environmental Issues:
Three Basic Tasks

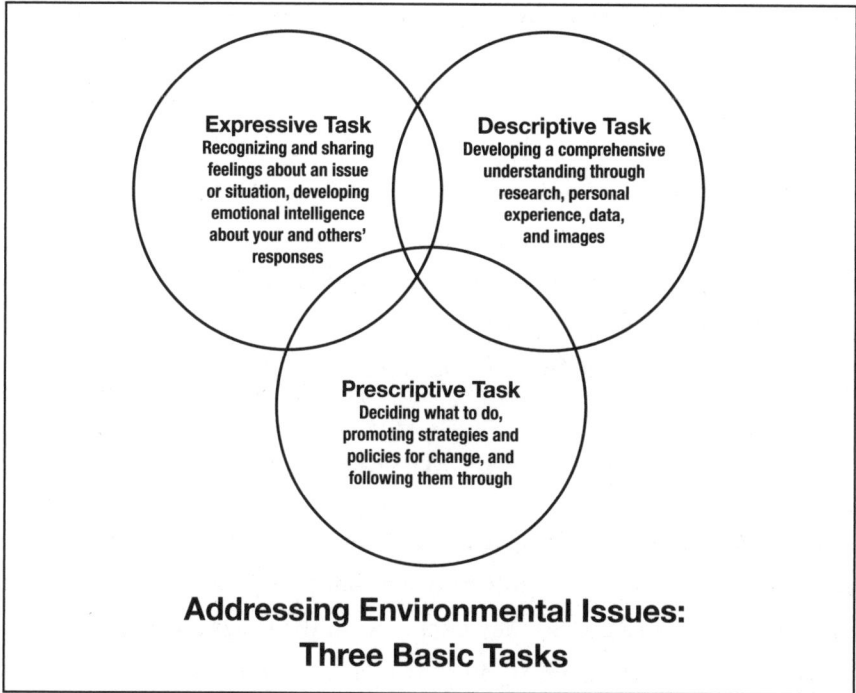

These tasks apply whether or not you are an expert. They occur simultaneously, and they are constantly evolving as you gather new experience.

Unfortunately, when environmental issues are addressed in the fields of economics, policy, law, and science, the spotlight is almost exclusively on the second and third tasks. Indeed, it often seems that these two are the *entire* focus of science and policy. There is no "how you should feel about this troubling material" section in scientific papers on climate (or in most news stories or classroom discussions of climate issues). As we have seen, if there is no recognition of or guidance for emotional expression, it will bubble up on its own, in the form of eco-anxiety or despair. Emotional expression is a needed counterbalance to the catalogue of climatic disasters people see on their social media and news feeds.

Put another way, the *emotional labor* of managing emotions involved in eco and climate action has been largely ignored, denied, or avoided. It's left to journalists, activists, young people, and the public at large (including you) to fill in the missing emotional pieces.

To have an emotionally intelligent response to climate change, you need to give attention to all three tasks—expression, description, and prescription. Your emotional and gut responses are a crucial first step. But they need support and a reality check from your intellectual understanding, and an outlet through your actions. That completes the cycle. Even if you understand the problem and have opinions on what needs to change, that doesn't necessarily mean you know how to express your feelings about it or respond to others on a feelings level. This can limit your ability to communicate and restrict access to your motivation and your coping skills.

EMOTIONS AND FEELINGS

I tell my clients that *emotions are in our bodies and feelings are the language we use to express them.* All sentient beings have basic emotional responses of approach or avoidance as they interact with the world around them and move toward or away from stimuli. Imagine a sea anemone in a tidepool and how it will quickly draw itself inward in response to your touch. For us humans, feelings words, in the context of our language and culture, are the doorway to recognizing and regulating our emotions. The wider your vocabulary to express your feelings, the better you'll be able to identify, articulate, and regulate your emotions.

I see parallels between the lack of expression of feelings associated with climate change and my early work with folks recovering from heart attacks and open-heart surgery. When I was a doctoral student, I focused on health psychology and behavioral medicine, and I studied the emotional process of patients at a cardiac rehabilitation clinic in rural New Hampshire. These were stoic, hardworking people, mostly men, blue-collar, rugged individualists who reminded me of my father and uncles in Buffalo, New York, where I grew up. They were happy to talk about even the most garish of heart attack stories in all their gory detail, but when it came time to describe any fear or anxiety about their health or mortality, they clammed up. They weren't used to self-expression and many exhibited what emotions researchers call *alexithymia* (lack of language for emotions). With no ready vocabulary to express their feelings, there was no emotional pathway to coping with their health problems or connecting with their concerned spouses and children. I remember one old gent joking, "Nobody knew I had a heart until I had a heart attack." It's important to respect that we all tend to avoid things we are not good at or don't feel competent doing. But *climate alexithymia* is not helpful, and becoming better at expressing emotions is a skill anyone can learn.

Delphina knew she desperately wanted to get better at talking about her feelings but didn't know where to start. Like the old adage goes about teaching someone to fish, I gave her and her fellow veterinary students a tool they could use in the long term: a feelings vocabulary list, which contains words for feelings, bodily sensations, and associated behaviors that flow from emotional states. For example, if you have a feeling of *apathy*, this might be accompanied by a bodily sense of lethargy and behaviors like isolating yourself and avoiding activities, even ones that make you feel better. Feeling *empowered* has an accompanying sense of energy, and makes you more likely to speak up for yourself and take action on projects. Using the list, anyone can learn to accurately describe their emotional experiences and actively move toward emotional health.

But even with the right vocabulary, you may still have some barriers to experiencing and expressing your feelings. I call some feelings "protected feelings" because they are so sensitive we protect them like a wound. We may not even want to admit to feeling them.

Some protected feelings Delphina identified were *guilt, fear,* and *loss.* Further, she identified the feelings that she *wanted* to feel: *courage, hope,* and *confidence.* It's a leap to go from guilt to hope or from fear to courage. I call those goals "stretch feelings." They are ambitious but absolutely possible. Since we can't just flip a switch and select a fresh feeling, the key is to work toward them gradually and find some stepping stones within your grasp. For example, Delphina identified middle-ground emotional states that felt more achievable and would transition her gradually toward her goal: awareness, presence, patience, curiosity, compassion, optimism, and mindfulness. If she couldn't yet feel *courage,* perhaps she could feel *curious.* If she lacked *confidence,* she could shoot for feeling *aware* and *patient. Hope* seemed unattainable, but feeling *optimistic* was within her reach. Even talking about positive feelings made her feel *energized*. That's the secret of positive emotions. As positive psychology researchers including Barbara Frederickson have demonstrated, positive emotions tend to build upon themselves and lead to a sense of opening. By asking herself what feelings she could inhabit now and what feelings she wanted to cultivate, Delphina activated a growth mindset and gained more control over her emotional life.

EXERCISE: Examining Your Feelings

As you consider climate change and the natural environment, ask yourself:

1. *What am I feeling?* Take time to identify all the specific feelings you have. You can use the examples in this chapter or in the Glossary at the end of the book.
2. *What do I want to feel?* If you have difficulty with protected feelings, what feelings can you inhabit right now that are stepping stones on the path?
3. *Which feelings might I stay with in service of my growth and wisdom?*

VARIETIES OF CLIMATE EMOTIONS

My podcast colleague, Finnish emotions researcher Panu Pihkala, has drawn upon decades of research to create a map of the many kinds of emotions people globally may feel about the current environmental crisis.

Emotions tend to differ on their valence (whether they are pleasant or unpleasant) and how strongly they are experienced. Scan through the categories below, and notice which emotions feel more common and accessible to you, which are sensitive and protected, and which are a stretch.

What feelings would you like to feel? Which ones do you need to feel? Imagine yourself being more comfortable holding all of these various feelings as they arise, expressing them to yourself and to others, and also letting them pass.

Negative or Unpleasant Feelings
- Anger (including frustration, irritation, and rage)
- Anxiety and fear (including powerlessness, terror, and panic)
- Confusion (including feeling overwhelmed, frazzled, or withdrawn)
- Depression (including sadness, meaninglessness, hopelessness, and numbness)
- Disgust (including aversion and resentment)
- Envy (including jealousy and resentment)
- Guilt and shame (including embarrassment, inadequacy, remorse, and self-loathing)
- Hostility (including skepticism, contempt, hate, and schadenfreude, or taking pleasure in another's misfortune or suffering)
- Indignation (including outrage and betrayal)
- Sadness (including grief, yearning, loneliness, and *solastalgia*, the sense of loss about disappearing nature)
- Surprise (including shock and disappointment)
- Worry (including dread, helplessness, and overwhelm)

Positive or Attractive Feelings
- Competence (including feelings of being capable, empowered, proud, and assertive)
- Empathy (including compassion, forgiveness, and grace)
- Gratitude (including feeling appreciative, sufficient, honored, or blessed)
- Inspiration (including feelings of insight, awe, elevation, and transcendence)
- Joy (including pleasure, happiness, amusement, and bliss)

- Love (including care and belonging)
- Motivation (including excitement, commitment, duty, and determination)
- Optimism (including hope, trust, faith, and confidence)
- Surprise (including wonder and amazement)

USING A GLOBAL VOCABULARY FOR GLOBAL PROBLEMS

The international veterinary students came to their environmental experiences with rich emotional vocabularies spanning many languages beyond English. One Spanish-speaking student taught us the word *agotado*, meaning "exhausted," "worn out," "emptied out," or, as she put it, "no drips left." The German *Weltschmerz* (literally "world pain") captures the feeling of carrying the weight of the world on your shoulders. The nuanced Korean word *han* means "sorrow and resentment while patiently waiting or hoping for a solution." The Portuguese word *saudade* refers to a longing, melancholy, nostalgia, or yearning for a happiness that once was or was possible: a fitting word to capture the bittersweet nature of solastalgia and environmental grief.

And there are ways to verbally express uplifting feelings about our relationship with nature that we can borrow from other languages, too. The Japanese feeling of *mono na aware* means appreciating the transiency of the world and its beauty. From Maori we have *turangawaewae*, literally "a place to stand," where we stand feeling empowered and connected; it refers to where we belong. From the Navajo, we have *hozho*, "balance and beauty," referring to harmonious relations between humans, nature, and spirit. From Finnish we have *sisu*, "extraordinary courage and determination." From Yiddish, there is *chutzpah*, which means to have cheek, nerve, confidence, and audacity. From Danish we have *arbejdsglæde*, literally "work gladness"; it refers to pleasure or happiness derived from work, and something to celebrate when you find a fulfilling task or mission, especially with a supportive group of compatriots. Whatever language you choose, the nuanced expression of these emotions can be quite beautiful.

HOW YOU EXPERIENCE ENVIRONMENTAL EMOTIONS

Of course, in terms of planetary-scale issues, you feel *many different emotions at once*. My colleague Panu also talks about *compound emotions* like hope, grief, and melancholy, which require holding both positive and negative feelings in mind—sort of like doing a yoga pose in which different limbs are held stretched in different directions. Just like yoga, holding different emotions at once takes practice.

Feelings can also be wild. Control is elusive no matter how hard we try to shape our feelings. Like breathing, emotions arise without conscious thought; they show up and reveal themselves in quiet moments, at times of great joy, and in times of stress (especially when we are bombarded by bad planetary news). And as with breathing, we can come into awareness of our emotions. By paying attention to them and assigning them words, we can become attuned to their rhythms and even influence their depth and pace; this is the emotional regulation we discussed above.

It's also helpful to think of emotions as a sequence, part of your stream of consciousness. For this, I use the image of a *train of emotions*. The train's engine is the primary emotion, which animates the others that follow. For example, surprise is often a hidden primary emotion regarding environmental issues, and can be negative or positive. You might be startled by a troubling news item or caught off guard by a remark made by a loved one or friend. In this case, surprise triggers a sense of vulnerability and threat, followed by secondary emotions like anger, sadness, or confusion. However, you might also be caught by surprise witnessing a beautiful sky or sunset, sighting a rare animal, or learning about some positive environmental news. In this case, surprise can lead to awe, gratitude, or contentment.

COPING WITH TROUBLING EMOTIONS

I was direct when validating Delphina, Bidisha, and Kamau: If you go into wildlife conservation with strong values and an open heart, as you need to do to fully experience the work, you will also get hurt. You will confront suffering, injustice, tragedy, ignorance, mendacity, absurdity. This dilemma was captured beautifully by the writer and conservationist Aldo Leopold, who wrote, "One of the penalties of an ecological education is that one lives alone in a world of wounds. Much of the damage inflicted on land is quite invisible to laymen. An ecologist must either harden his shell and make believe that the consequences of science are none of his business, or he must be the doctor who sees the marks of death in a community that believes itself well, and does not want to be told otherwise."[2] Yes, your heart can be broken—and it can be mended.

To heal these invisible wounds, it's helpful to identify your preferred emotional coping style. We all toggle between two basic emotional styles about eco and climate feelings: needing to embrace the dark and needing to embrace the light. These are exemplified by the *despair and empowerment* approaches used in environmental activism and the *broaden and build* approaches that come from the positive psychology tradition.[3] While both are useful and necessary depending on the context, you may tend toward or feel more comfortable engaging with one or the other.

In a despair and empowerment approach, you go deep into your feelings, however painful, and learn to express them in a supportive social group. You metabolize grief, pain, loss, and anger, and this provides a release, a catharsis, a clearer sense of purpose. The vet students used the phrase "vomiting emotions," and for good reason. Catharsis is when you can let go of big feelings and feel an emotional release, especially of visceral emotions such as grief and anger. (The term comes from the Greek word *katharsis*, which means "purification" or "cleansing.") You are empowered by walking through the fire of feelings.

Broaden and build approaches take the perspective that positive feelings breed more positive feelings. The way you mend and empower yourself is to connect with beauty, your love of nature, and feelings like curiosity, insight, gratitude, and awe.

Ultimately, you need to develop a growth mindset for experiencing the light *and* the dark.

Paradoxically, to endure and find meaning in negative emotions you need to make sure you are open to positive ones. The veterinary students expressed a sense of being stuck on a limited range of feelings about their research—sad, frustrated, angry, powerless, grieving. It's easy for this to happen, especially since the majority of climate news we see is negative. In order to get unstuck, I asked them to imagine holding an old-fashioned magnetic compass. No matter which direction you turn, the needle swings to point north. But if a compass needle is stuck, it points in one direction no matter how you turn. Your emotional life can feel stuck too.

The goal is to practice having a working 360-degree *emotional compass* accessible to you that points at whatever eco or climate emotion is appropriate for the moment. If the view outside your door is pleasant and calm, you deserve to feel pleasant and calm. If you come upon a beautiful nature scene or hear an uplifting speech, you want to be able to feel appreciation and inspiration. If you witness something troubling, it is appropriate to be sad or afraid.

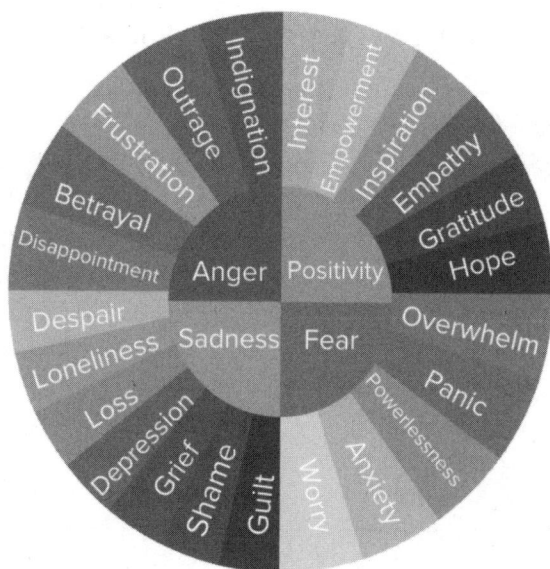

A Climate Emotions Wheel Based on Panu Pihkala's Research
Source: "The Emotions Wheel," CIRES Center for Education, Engagement, and Evaluation, University of Colorado, 2024, https://ceee.colorado.edu/resources/emotions-wheel.

It's important that you don't try to channel all of your emotional needs through a single emotion or a limited range of emotions. Some emotions, like presence, curiosity, patience, compassion, satisfaction, awareness, contentment, or just "chilling," are generalists, appropriate in a range of settings, both comfortable and uncomfortable. Conversely, emotions like anxiety, anger, grief, joy, empowerment, and awe are specialists, meant to play specific roles in your functioning and survival. Anxiety, for example, is devoted to self-protection and attention to possible threats.

To illustrate this for the vet students, I used the metaphor of a *climate emotions party*. If you only invite one emotion (say despair), it will be a pretty grim affair—not unlike a birthday party with just one guest. But if you invite other emotions to join despair, like awareness, compassion, love, humor, and courage, you'll have a fuller experience. You can't uninvite your emotional responses. They all play a healthy role and will crash your party unbidden—but you can always add others to the emotions guest list!

GROUP EXERCISE: Acting Out Climate Emotions

Write down various emotion words on small pieces of paper, fold them up, and put them in a basket or hat. Then pass the basket or hat around and ask everyone to pick out an emotion and act it out without words, while the others guess what emotion they're expressing, like a game of charades (you can do this in an online group by assigning words to people via private chat). People can get creative by adding facial expressions, body postures, and sounds. When the group guesses correctly, they should validate the person's expression with a supportive response like "We feel you."

There will be times when you just feel *numb*. This might be a normal instance of alexithymia, like the lack of language for emotions I found with the cardiac patients; if so, you can address this by consulting a feelings vocabulary list. Numbness might guard a sensitive or protected feeling or indicate a reluctance to commit to a stretch feeling for fear you can never reach it. Numbness may be a sign of deep fatigue or burnout, manifesting as the anhedonia—inability to feel pleasure in typically pleasurable things— that Delphina experienced. And in more serious cases, emotional numbing and anhedonia can also arise in people who have lived through extreme personal traumas or disasters that disrupt their emotional responses and memories of past events, as in the case of post-traumatic stress disorder. I will share more advanced techniques about how to work with issues of trauma and depression in Part Three.

Sometimes, when you are feeling overwhelmed, you need to know the difference between a state of burnout and simply feeling depleted and needing rest. I reminded Delphina that, often, *despair is fatigue in disguise*. Personally, I know that when I am frustrated, irritable, or anxious, it's a cue that I need to take care of myself. The importance of rest and self-care takes center stage as you learn to develop personal sustainability in the next chapter.

Chapter 3

Calming

The territory is our body.

—Ana María Hernández Cárdenas[1]

IF YOU CAN, RIGHT NOW, WHETHER YOU ARE sitting or standing, listening or reading, be aware of your body. Take a conscious breath. Fill your lungs, then slowly exhale. Feel gravity flowing down through you. Trust your rigid skeleton to hold you up, and let your muscles relax and hang from your bones. Roll your neck slightly and let it loosen. Notice how your heavy skull balances lightly on the top of your spine. Consider the beating of your heart, the circulation of your blood, the faint electric pulses of your nervous system. Imagine the communication between different structures of your brain: the primal brainstem impulses that regulate your body temperature and breathing; the ball of your cerebellum at the base of your skull helping orchestrate your fine movements; the amygdala in its bunker deep in your limbic region, parsing your senses and scanning for threats; the frontal cortex, a busy middle manager for your responsibilities and desires.

This is a moment of being embodied. And that is the goal of this chapter: learning to *be in your body*, aware in the present moment, feeling all its sensations—emotional and physical—as you cope with climate anxiety and its stresses and opportunities. When it comes to the mental health impacts of climate change, *the body keeps the score.* Your body is also a source of joy and connection. On a fundamental level, your physical body is a part of Earth's ecology. Your physical health is connected to the health of the environment. Thus, working for environmental sustainability rightfully includes your *personal* sustainability.

Personal sustainability includes taking what steps you can to live a balanced life, support your mental health and well-being, and reverse unhealthy habits that deplete your energy and compromise your emotional resilience. The better you feel, the better you can protect your community and place, and expand sustainability outward.

PERSONAL SUSTAINABILITY

A classic definition of environmental sustainability is "meeting the needs of the present without compromising the ability of future generations to meet their own needs."[2] If you apply this sustainability ideal to your own daily life, it makes sense to attend to your basic human needs like rest, healthy diet, exercise, and social support. Yes, there are times you will need to make sacrifices and put your needs second to the needs of others—as the classic maxim goes, "to live simply so that others may simply live." But in general, it makes no sense to endanger your health to save the planet.

This idea of personal sustainability first came to me a decade ago when I took time away from my daily life to journey through Wyoming's snow-covered mountains with a friend who had worked with me as a wilderness guide in our youth and was now doing research tracking the elusive wolverine populations. Wolverines depend on dens buried in a deep, lingering snowpack to nurture their new cubs. Like the ostrich mother that inspired Larysa, wolverine mothers' tending is under threat due to the dwindling winters of climate change. One cold night, sitting near the woodstove in an old trapper's cabin, my friend asked me about my life in the city. The truth was it was hard. I was disconnected from the natural world in a way that didn't feel comfortable. That's when I realized that I'd been spending a lot of time thinking about environmental sustainability and very little time sustaining myself. On that wilderness retreat, the lessons from the mindfulness practices and cardiac rehabilitation I had studied during my psychology training, my studies of psychology and modern identity, and the need to address the rising stress over our ecological and climate crises came together for me. Once I got back home to the city I made a commitment to practice mental and physical health as a form of *personal sustainability*—in my life and as a psychologist and therapist.

Putting sustainability in action in real life is another story. Even with the best of intentions, it can be difficult to balance all the roles and duties we expect of ourselves in modern society. In his classic 1998 study, *In Over Our Heads: The Mental Demands of Modern Life*, developmental psychologist Robert Kegan portrayed modern life as a curriculum with many subjects to master: You need to be personally accomplished, financially secure, smart, fit, good at your job, a compassionate partner, a wise parent, and a conscientious citizen.[3] If that's not enough, now in the twenty-first century you have new courses in your life curriculum, including managing media saturation, and a perception of out-of-control political and climate crises.

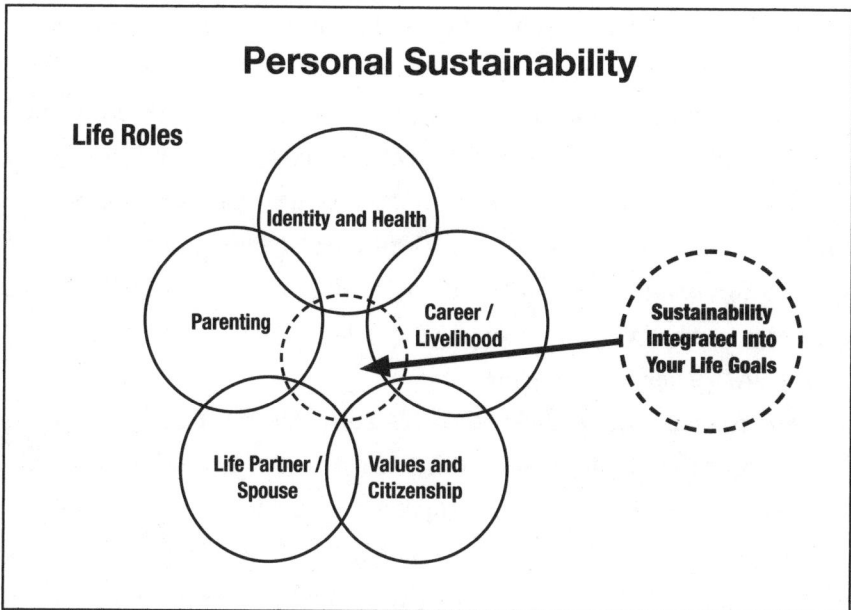

Personal Sustainability

Life Roles

Identity and Health

Parenting

Career / Livelihood

Sustainability Integrated into Your Life Goals

Life Partner / Spouse

Values and Citizenship

Integrating Sustainability into Your Life Roles

Dear reader, it's normal and natural for us to feel *in over our heads*. The key is to learn how to swim.

Balancing all this requires metacognition (the "thinking about thinking" I discussed in Chapter 1) and being able to express and help guide your emotions (in the form of emotional intelligence, as introduced in Chapter 2). It also requires some ability to *reclaim your nervous system* and balance the demands of the big picture with your day-to-day well-being. Rather than thinking of working toward a safe and healthy planet as a whole new life course to manage—as if this were *separate* from your roles as person, partner, parent, worker, and citizen—think of personal and environmental sustainability as being *integrated* into all of your life roles.

Marcus

I first knew Marcus from afar as a charismatic speaker at "Smart Cities" urban planning meetings in Portland, where he worked for the city office of sustainability. In late 2020, Marcus came to see me as a client, partly at the urging of his wife, who hoped that he would talk to someone about his unpredictable moods and high stress levels. Marcus had not seen a counselor before, but given my presence at city meetings and reputation for addressing climate angst, he sought me out.

Marcus and I had much in common. Our kids attended the same public school. We both grew up in blue-collar families and were the first in our families to go to college. We were both Generation Xers and bonded over shared interests in obscure 1980s hip-hop and underground music, the sports teams of our childhoods, and progressive politics. As we approached our fifties, our bodies were becoming more like we remembered our fathers' to be than like ours as young men.

And we had obvious differences. Marcus is Black and I'm white. I was a newcomer to Portland and he was born here. I migrated to the Pacific Northwest for its ancient forests, rugged coasts, and open-minded cities. Marcus's family came seeking jobs and opportunities during World War II, when the ship and warplane factories made the Pacific Northwest a terminus of the great African American migration. The areas in Portland that

I was drawn to for their funky, historic character were formerly the main streets of Black neighborhoods, isolated by racist real-estate "redlining" policies that kept people of color from obtaining mortgages and decimated by urban renewal programs that bulldozed poor neighborhoods to clear the way for expressways and sports arenas. By the time I arrived in the city, baristas, boutiques, and bike lanes filled the vacuum.

Despite our similar educations and tastes, our different racial and cultural backgrounds led us to have different environmental identities. For me as a white person, to opt out and head West to live the life of a dirtbag river rafter and environmentalist was a choice and privilege. For Marcus as a Black person, marginalization was a given, and ecological concerns a necessity; his capital-*I* issues focused on protecting his home place. Because he was a city planner in an era of deadly heat waves, his connection with nature was less about the Northwest's old-growth forests and more about getting shade trees and air-conditioning into low-income neighborhoods that had become urban heat islands. Embodying his environmental values meant bearing witness to history. For him, the Columbia River was less a place to go rafting than a reminder of environmental injustices like the infamous 1948 Vanport Flood, which wiped away a predominantly Black community of 18,000 shipyard workers who were settled in a known floodplain.

Marcus's personal-level little-*i* issues focused on his physical health. His father had suffered strokes and died of heart failure. His older brother had diabetes. Despite being an elite college athlete in his younger years, Marcus now had high blood pressure and was himself at risk for heart disease. Emotionally, Marcus was also what therapists call an "internalizer": someone who keeps their stress inside. In social settings, Marcus was an upbeat presence, someone who always carried a hint of a smile on his lips. But in meetings at city hall, when listening intently, or when giving a careful response to a complex question about the needs of Black residents, Marcus's gaze over his eyeglasses could be intense, approaching a scowl.

As pressure at work mounted, Marcus became distracted and emotionally distant from his family. Most nights, his scowl followed him home.

When he caught his own reflection, it reminded him of times his father had been bitter and judgmental. Between the sense of duty and protectiveness he carried about the livability of his community (particularly the plight of the unhoused), a sense of responsibility for his immediate and extended family, his own health worries, and now tension with his wife and kids, Marcus's personal ecosystem was stressed beyond a sustainable level.

FIGHT, FLIGHT, FREEZE

In reclaiming your nervous system, you must respect your body's evolved responses. When you perceive an intense environmental threat, whether by taking in real-time news images of an actual disaster or through your imagination, your survival response takes hold. Your brain's hypothalamus instantly activates your sympathetic nervous system, triggering the release of adrenaline, epinephrine, and the stress hormone cortisol. Your heart pumps faster. Your muscles tense, ready for action. Your blood pressure spikes and your digestive processes shut down. Your liver sends out glucose to give your body energy. Your breathing becomes rapid and shallow, and your skin cools as blood is shunted to your major muscles and vital organs. Your eyes dilate. Your senses become acute. Even your blood becomes thicker, more likely to coagulate in anticipation of a potential injury. At the body level this activation of our sympathetic nervous system is your body and mind's involuntary reaction to danger, the classic *fight-or-flight response* described by physiologist Walter Cannon in 1915. It's a beautiful mechanism honed by evolution for survival. Like a deer standing still to blend into the forest until a predator passes, humans may also freeze in an adaptive response to extreme situations of disaster or violence. The immensity of global threats— like the ones Marcus confronted at work on a daily basis—can certainly trigger any of our threat responses.

Marcus was unconsciously bringing his activated survival response home, so even small interactions with his wife or kids led him to feel emotionally flooded and unable to communicate effectively.[4]

But the sympathetic response of fight, flight, or freeze can be tempered by our parasympathetic nervous system, which allows us to *rest and digest* under stressful conditions. That ability is what underlies humans' ability to achieve under pressure, like a basketball player calmly launching a free throw in a raucous sports arena. To understand how these two aspects of our nervous system coexist, clasp your two hands together in front of you and let them push against each other. Then let one of your hands go limp. The dominant hand is like your sympathetic nervous system. In a wrestling match, the sympathetic usually comes out on top—your body and mind are keenly and innately aware of threats, and therefore are easily triggered. Moreover, your sympathetic response is typically much stronger from being exercised in your daily life, jolted by every bad news story.

A key tool for coping with eco-stress is developing a stronger relaxation response. Clasp your hands again and let the weaker hand (your parasympathetic, relaxation response) become stronger so that it balances the tension with the other hand and can itself become dominant. The solution here is not to try to eliminate stressors. You can't. The solution is to develop a relaxation response that can overpower your stress response.

Marcus's sympathetic nervous system was not getting enough push-back from his parasympathetic system. He needed to find a sustainable balance.

THE UPSIDE-DOWN PYRAMID

There is an exercise I developed during the terrible summer of 2020 in Portland when my clients and I were contending with COVID restrictions, threatening wildfires, unprecedented heat waves, and social unrest that captured worldwide attention. We felt overwhelmed and inadequate in the face of all these novel threats on top of our daily hassles: everyday strains about money, health, relationships, and dealing with issues like noise, pollution, and crime. Not only were people's minds hijacked by anxiety, but their bodies also were on high alert.

All the Issues and Challenges Weighing on Me

Issues and Challenges

Perception of Inadequate Resources
Feelings of Stress, Anxiety, and Despair

Daily Actions for My Well-being and Values

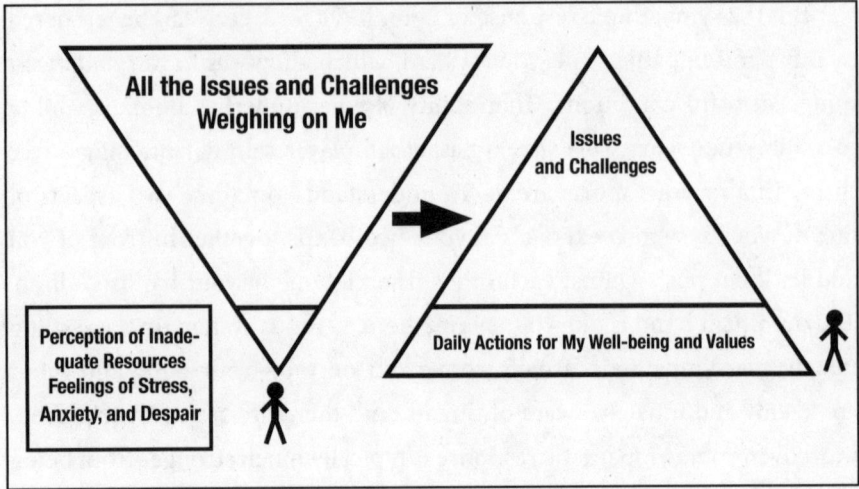

When Marcus's pyramid was upside down, the challenges at the wide top seemed infinite, and he felt them weighing him down. A key realization for Marcus was that if he turned his pyramid right side up and took care of his foundation, it could withstand the weight of the capital-*I* issues causing him stress—climate change, social justice, caring for his family, serving his community. I wanted him to tackle these too, but I wanted him to have a secure base to do this *sustainably*, over time.

EXERCISE: Drawing Your Pyramid

1. Draw a triangle to represent a pyramid, but pointing upside down, with a little stick figure underneath. This figure represents you, a reluctant Atlas, bearing the stress of the world. The wide top is filled with all your stressors. Your foundation becomes reduced to the downward tip of the upside-down pyramid—a meager triangle of personal resources. There are times when the upside-down pyramid seems to expand infinitely upward with all the things you worry about. It feels crushing—and totally unsustainable.

2. Now draw a new pyramid that is right side up, steady and firm on a wide base. Your stressors are still there crowding the tip

of the pyramid. But this time the foundation of the pyramid is wide enough to accommodate all the things that you do every day to be your best self. Some of the bricks that make up your foundation are universal: We all need water, food, rest, exercise, and supportive relationships. Most of us have daily work. And we have resources unique to us: our relationships with our places and communities, self-care, our hobbies and passions, and our sense of meaning or spirituality.

FATIGUE IN DISGUISE

Marcus recognized that unwelcome feelings and sensations were making him irritable and distancing him from his family: a case of *despair is fatigue in disguise.* So we identified the high-risk zones in his daily life and came up with actions he could take to prevent himself from starting his day with internalized stress. Then we added a practice to help him be more present and open with his family. Here are some of the bricks in Marcus's foundation:

- *Healthy morning ritual.* Marcus instituted a new morning routine that would allow him to begin the day with human and nature connections, not screens. First thing in the morning, before he looked at his phone and email, he would take a morning walk outside, just around the block, making a point of greeting his neighbors.
- *Rest.* When life is moving quickly, don't underestimate the virtue of learning to pause. Marcus was familiar with the Nap Ministry, a movement sparked by Tricia Hersey to encourage Black activists to claim radical rest, or self-care. In the spirit of Hersey's declaration that "Rest is resistance," Marcus incorporated rest among the bricks in his foundation: brief moments when he could stop, take calming breaths, and refocus his mind on his big-picture goals.

- *Healthy diet.* Marcus loved food, from the farm-to-table fine dining of Portland to backyard barbecues. Being mindful about his diet included being careful about not eating unhealthy food when he was stressed and also balancing the ethics of what he ate, including how it was produced and by whom, with his tastes and enjoyment. (See Marcus's Food and Diet Values Chart, below.)

- *Gratitude.* Research shows that cultivating one's attitude of gratitude improves emotional well-being. Marcus made space in his day to give thanks to God for his life, his health, his family, and his good fortune to have these when many others did not.

- *Change the channel.* This brick is a great example of what I mean when I tell clients, "Your life is the news." Marcus's new habit of walking around the block each morning was an opportunity to get a dose of neighborhood news, including noticing the bird life flitting through the trees along his block, before he turned his attention to the news in the headlines. During the day, when he had a craving for media news he substituted radical rest.

- *Evening transition ritual.* I recommended Marcus add a transition time when he arrived home from his job in the evening: a few minutes to pause, give thanks, and reimagine his pyramid to ensure his family was at the top of his priorities. It can be hard to resist answering work emails in the evening. Creating a conscious habit of not even looking at his inbox until the next business day was a healthy boundary.

It can also help to have an end-of-the-workday mantra. Marcus's was "Work to home. Mind to body. Body to place. Place to people. People to heart. Heart to mind. Mind to body."

There are times when, like Marcus, *you* will feel upside down—overwhelmed, inadequate, and desperate, struggling under the weight of your greatest concerns. A gift of the pyramid imagery is that it helps you see how your actions matter. They matter for you, if nothing else. And if you see your *self*-system as a deserving part of the larger *eco*-system, then by helping yourself to be healthy, you help the planet to be healthy too.

Take in another breath, this time as fully as you can. Feel the air in your lungs like two balloons stretching your rib cage. Now slowly exhale as completely as possible, compressing the muscles in your chest like you're squeezing all the water out of a sponge. Then gently relax your belly; feel your lungs open and air rushing in to fill the temporary vacuum you created. This is a mindful breath, and a small step toward reclaiming your nervous system.

EXERCISE: Flipping Your Pyramid

1. Draw a sketch of your pyramid when it is upside down. What overwhelms you? Then right the pyramid and identify some key bricks in your foundation. Focusing on your foundation creates the health and energy you need to take on your big capital-*I* goals over time. The top of the pyramid becomes an arrow pointing at your goals. The arrow shape challenges you to prioritize what is on top and to be more decisive and efficient in how you spend your precious life energy.

2. Keep your sketch by your bedside and ask yourself before going to bed at night, "Have I hit on all my bricks?" If so, that's a good day. You can't always hit on all your bricks, and that's okay. But if you start to see a pattern of missing bricks day after day, make a note of this.

ECO-STRESS MANAGEMENT TOOLS

Like many people, Marcus wasn't conscious of the high baseline of physiological arousal he carried. It was easy for one upsetting news story or one traffic jam on the way to a meeting to be the final straw that sent him into the stress danger zone. To help him gain more awareness, I had Marcus imagine an old-fashioned mercury thermometer to represent his stress level.

As your stress level rises your senses become more focused and you see the world through a narrow lens. This is very useful during an emergency. If you are stuck in a burning building, it's not helpful to stop and observe the art on the walls. But chronic daily stress leads to tunnel vision and rigid thinking, which are counterproductive when you are trying to solve complex problems or collaborate with coworkers and family. By lowering your stress, you gain bandwidth to be more aware, creative, and psychologically flexible. Marcus checked in with himself throughout the day. If his stress level was too high, it was time for active measures like radical rest to bring the level down.

I also called on an old cardiac rehab tool to help Marcus fend off ecological stressors: the Hook, a simple exercise developed to help heart attack survivors alter the hard-driving, type A behavior traits that are often associated with coronary artery disease.[5] High-achieving people trying to solve urgent environmental problems can fall prey to these type A tendencies, particularly when they are fatigued or face setbacks. The Hook helped Marcus shine a light on the eco-stress triggers, or "hooks," in his daily environment and be more conscious of how he chose to react to them.

The exercise uses a simple visual: an image of a fishhook with a lightbulb for insight next to it. Like a fish, we tend to get snagged again and again on the same stressors. As a fly fisherman, Marcus knew that successful anglers are those best able to trick a fish into biting by using well-disguised hooks. This led him to the realization that many of his stress hooks were also camouflaged or hidden, and drew him in instinctively. This was a key insight for him. We can't help but be drawn to engage with environmental issues that are important to us. The goal is to act mindfully. We may still get caught unawares at times, but as the old-timers in the cardiac unit in New Hampshire used to say: "If you are on the lookout, you can keep the hook out."

Personal Sustainability:
Balancing Values About Food and Choices

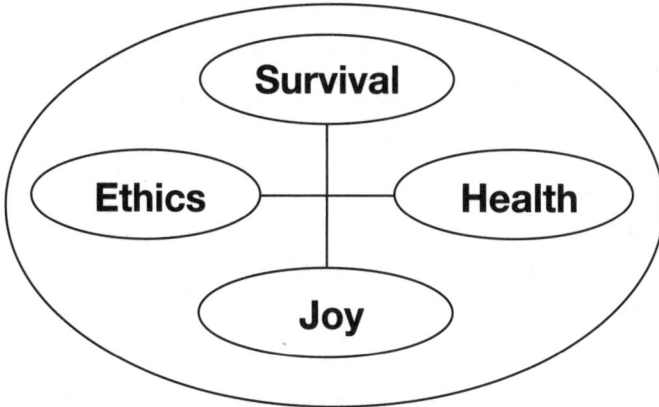

Survival

Ethics — Health

Joy

Marcus's Food and Diet Values Chart

Earlier, we saw how Marcus was making heart-healthy diet choices a foundation of his sustainability pyramid. Consider how food figures into your own health and values. While access to food can be taken for granted, it's important to realize that for many people around the planet basic nutrition and survival are not secure and are the main priority. Once you have access to food, especially a range of options, the challenge is to balance your physical health and fitness needs both with your ethics regarding food production, animal suffering, and the environmental impacts of industrial agriculture and meat production and with your instinctive striving for enjoyment and pleasure. Marcus was attentive to the social justice aspects of the food system, including the rights and safety of frontline workers in agriculture and the meat industry. He was also inspired by efforts to reclaim African American history and autonomy in terms of farming and food production.[6]

TAKING CARE OF YOUR ANCESTRAL BODY

To reclaim your nervous system in the era of climate change, it's helpful to respect how much of the constantly changing, technology-heavy, and consumer-focused modern world is new and different from the environment in which humans' bodies and minds evolved.

There is a classic 1950s television comedy routine from *I Love Lucy*. Lucy and her friend Ethel go to work at a candy factory, and their job is wrapping chocolates that pass by on a conveyor belt. Things go well until the conveyor speeds up. Lucy and Ethel struggle, sloppily wrapping some chocolates and missing others. As the belt moves ever faster, they become desperate to keep up, shoving candies down their blouses, under their hats, and ultimately filling their mouths. This skit spoofs the famous assembly line scene in Charlie Chaplin's 1936 film *Modern Times*. But where the trickster Chaplin ultimately rebels and sabotages the factory, Lucy and Ethel's response is more like how most of us respond to the accelerating speeds of life's conveyor belt: We scramble, and try our best to keep up. We act normal and hide our struggle. We stuff stress down, sometimes literally, in the form of comfort food.

I once took part in an exercise that evoked human evolutionary time, led by pioneering eco-psychologist Chellis Glendinning.[7] If you walk backward along a line thirty feet long that represents the 200,000-year history of modern humans, the world we know is barely a millimeter. The earliest near-instantaneous communications technologies that operated in any way like ours (for example, the telegraph) are only about 180 years old, dating from the 1840s. After the first eight inches, we leave behind the last 6,000 years and the history of civilization. The rest of the line takes us back into deep generations of human ancestors, all living in diverse ways that were closer to the land, gathering, planting, and hunting. Yet we share the same eyes, hands, brains, hearts, muscles, stomachs, and uteri as those people from long ago. We all drink, eat, defecate, sweat, dream, get pregnant, and have babies just like them. We all have our social groups, we all get embarrassed, we get lonely, we care and feel cared for, we laugh and cry. The basic needs of our bodies and minds are the same as they were for our ancestors—healthy food, vigorous exercise, community, family, and creative expression. But these early ancestors never experienced anything resembling the information technology that has evolved at a breakneck pace in the last few decades, and we all need to adapt.

One way to honor your body's natural wiring is to put guardrails on the amount of information you consume, not just news but also social media, emails, and documentaries. Humans are knowledge seekers and tool makers. Cellphones did not magically appear. We evolved to create them. The mismatch between technology and health, like Lucy and Ethel's candy conveyor, is not magical either. Just as we are hostages to a system that enables the climate crisis, our nervous systems are also hostages to the capitalist forces driving the information economy.

MANAGING YOUR NEWS AND INFORMATION DIET

This was certainly the case for Larysa, the postpartum mom anxious about disaster preparedness. Her pyramid was definitely upside down, and all she could feel was the weight of disasters and emergencies. Her foundation of coping tools had narrowed to a point and needed to be broadened into a wide base of support. Before her baby was born Larysa had practiced yoga, worked out in a gym, spent time with friends, and created art. These were important routines that she once fit in around her workday and family commitments. Using the image of the upside-down pyramid, Larysa became conscious of how much of her daily self-tending she was now neglecting. She needed to restore the bricks of her foundation—by going for a run, sitting down with a sketch pad, calling a friend. Elevating the goal of self-care led her to ask for help from her husband, parents, and siblings to get the time she needed for herself. In addition to her personal sustainability, she had to get wise about her stress hooks, which included scrolling through troubling news, especially in the middle of the night, when her mind leaped wildly to the worst conclusions.

Most of us have a choice regarding the media we consume, and here's where Larysa needed to do more work. Her screen-based electronic news intake was unbalanced: It went deep into the details of problems but rarely if ever showed solutions. It's important to note here that most digital news is served up to you by an algorithm that deliberately prefers dramatic negative stories—the more alarming the better—in order to capture eyeballs. This is why, as my colleague hope researcher Elin Kelsey estimates, only about 3 percent of news we receive about climate change is positive.[8]

However, news can be solution-focused rather than doom-focused. For a week, I had Larysa augment her typical daytime news intake with solution-focused journalism, which focuses on responses to a social problem, along with balanced, credible coverage of the solution's effectiveness and short-comings. Scrolling through the wealth of credible and positive stories on the Solutions Journalism Network—"How a Methodist Preacher Became a Champion for Black-Led Sustainable Agriculture"; "Denver's E-Bike Rebates Are So Hot They're Gone Within Minutes"; "Record Heat Waves and Droughts Can't Dry Up This Native Garden in Phoenix"—was a boost to her mood.

Larysa's psyche was habituated to predominantly negative, hopeless news stories. At first it seemed jarring, even artificial, to entertain actual positive developments in the world. When this happened, I guided Larysa back to her feelings vocabulary practice: She was beginning to feel *curious* and *aware* of good news, and hoped to eventually feel *trusting* and *confident* in what she was reading and learning.

Are we facing huge problems? Indeed. But if only 3 percent of your eco and climate news diet consists of positive events, you are not getting an accurate or balanced view of the world. Here are some additional ways you can add more personal sustainability to your news consumption habits. I find these practices help me to feel *more informed*, and in turn more present, grounded, and focused.

Making Your News Intake More Personally Sustainable

- *Analog news.* Access news through physical newspapers, magazines, or books. Pay attention to how this feels in your body. Note differences compared to screen-based offerings.

- *Avoid editorials.* Most "news" is actually just commentary or a one-sided take on a complex subject. Resist the compulsive urge to read stories that echo your beliefs and fears or that you expect will irritate you.
- *Gather news by ear.* For a day, just pay attention to news you hear spoken about around you. Notice how information is shared. Are people approaching topics using a fixed mindset or a growth mindset? How does the emotional tone people use affect you?
- *Practice regular news fasts.* Take a break from all news, analog and digital. Start with just a few minutes, then work your way up to several hours, to a day, to a few days.
- *Focus on your life as the news.* Be an observer of the news right outside your door: your people, community, and the natural world around you. Notice how this feels in your body and mind, and how it affects your daily pace, breathing, and sense of sustainability.

With all the stresses and responsibilities we bear and the information we expect ourselves to carry, is it any wonder that we feel in over our heads? Remember that taking care of yourself is an essential part of making the world a better place. Environmental sustainability is at the peak of your pyramid. Your personal sustainability is the base. The bigger the wicked problems you want to take on, the bigger the base of personal sustainability you need.

Chapter 4

Adapting

Stewardship means, for most of us, find your place on the planet, dig in, and take responsibility from there.

—GARY SNYDER[1]

CLIMATE ADAPTATION, IN PLAIN TERMS, IS CREATING A good life in the face of unprecedented environmental changes, learning to live with the changing weather and disaster threats that may beset you, and also finding positive opportunities within the crisis, such as making choices to be a healthier person and to live more in concert with your values. In 2009, as the clinical psychologist serving on the American Psychological Association's Climate Change Task Force, my task was identifying the "psychological impacts" of this looming issue. I want to help you understand some key lessons we learned in that research that have stood the test of time—and to help you use them to create your own adaptation and coping plan.

Every few years, the Intergovernmental Panel on Climate Change (IPCC) issues ambitious new reports that synthesize the latest scientific facts about how climate disruptions are affecting the planet. In this chapter, your task is to create a *personal* version of an IPCC report: your own "Individual Problems with Climate Change" report. The goal of this exercise is to clearly and objectively identify the specific climate and environmental dangers that affect your mental health and well-being in your particular region and community—without either catastrophizing or falling prey to disinformation and denial. We'll then add a flexible time frame for action, including actions you can take right now, in the near future, and in the longer term, as the climate situation continues to change.

We'll focus on actions to take during an actual disaster in Chapter 20. Right now, we are focused on building the mental and emotional skills you need to anticipate, proactively cope with, and bounce back from the dangers coming your way.

Reid

Reid, a teacher in southeast Texas, and his wife, Marisol, a nurse from Louisiana, are no strangers to bad weather. Reid grew up on the Gulf Coast and could name storms and hurricanes that were part of his family's lore going back for generations. He saw his parents lose their home to Hurricane Harvey, a Category 4 hurricane that made landfall in Texas and Louisiana in 2017, causing catastrophic flooding and more than one hundred deaths. Marisol was a teen when her Honduran family witnessed the destruction of Hurricane Katrina from their home in the Barrio Lempira neighborhood of New Orleans.

Reid and Marisol had two young children, whom they wanted to raise near Reid's family in Houston and Beaumont. Many people in Reid's family had worked in the local oil and fishing industries, and Reid always expected to stay in the region. He had always felt proud of having been born and raised in "hurricane country," but his sense of resilience had been shattered by the severity of recent disasters. He strongly identified with his region and his family, but now that he had a family of his own, he didn't feel safe staying in the place he'd always called home. He faced major life decisions and needed help making sense of all the factors in play.

To help Reid, I turned to my research, which identified three distinct ways climate disruptions negatively impact mental health:

- *Direct impacts.* The physical, psychological, economic, and political effects of disasters and extreme events. Disasters can take the form of an emergency (a single catastrophic event) or an ongoing crisis (a chronic problem that has reached a tipping point). These impacts are easy to see and understand: damage to and destruction of your local landscape, home, and property; loss of power, water, and community services; injuries and deaths; social breakdown; and psychological trauma.

- *Indirect impacts.* Long-term, post-disaster effects like storm recovery and rebuilding; supporting survivors and refugees, or becoming one yourself; disruption of the local economy and supply chains; and even famine, civil unrest, and armed conflict.
- *Emotional impacts.* The negative effects on your mood, outlook, and well-being from the emotional weight of global climate breakdown, the compounding effects of each new disaster, and vicarious trauma when reading about or witnessing the suffering of humans and other species and the destruction of cherished places—even when you are not physically or materially in harm's way.

The first two categories, direct and indirect impacts, have long been explored in the fields of emergency management and disaster studies. In human terms, a disaster is a serious and often sudden problem that causes extreme and widespread loss and injury, exceeding the ability of individuals, affected communities, governments, or whole societies to cope using their existing resources. In my hometown, Buffalo, New York, the difference between a snowstorm and a disaster is that the first means more shoveling and perhaps a temporary loss of power, and the second means you are looking for survivors trapped in their houses and bodies buried in cars. Indirect climate impacts are the shock waves: large-scale community and societal disruptions that often continue long after the original catastrophe is past, and the chronic environmental stress stemming from a continual pattern of disasters.

What was new in my research was bringing emotional impacts to the forefront as a central issue and showing that modern technology allows us to viscerally experience faraway traumatic events quite personally in our bodies and minds. The emotional impacts of a disaster can be significant, even when we are neither directly nor indirectly impacted by the event itself. These emotional impacts are now widely accepted and recognized by such terms as "eco-anxiety," "solastalgia" (loss and unease as your local environment becomes degraded), and "climate grief."

On a planetary scale, the overlapping ripple effects from the impacts of many local disasters, along with other social problems, along with our increased abilities to observe their effects and interactions, leads to a perception of *polycrisis*—multiple, simultaneous wicked problems.[2]

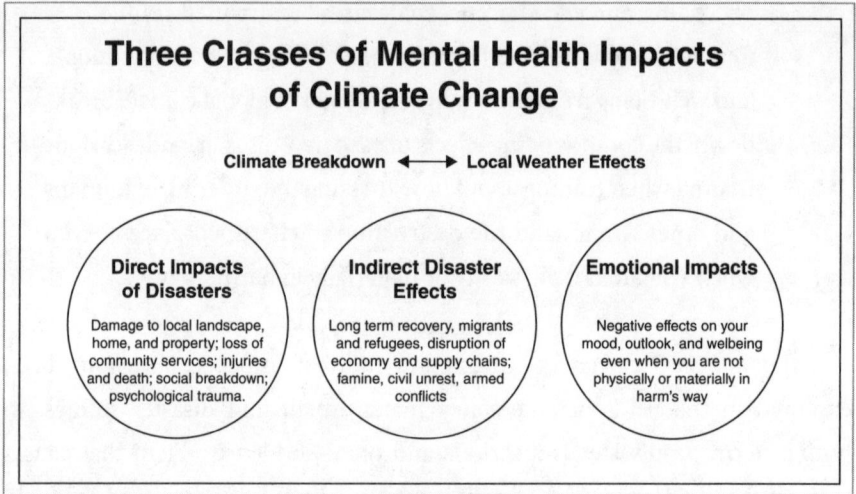

Three Classes of Mental Health Impacts of Climate Change

Climate Breakdown ◄──────► Local Weather Effects

Direct Impacts of Disasters

Damage to local landscape, home, and property; loss of community services; injuries and death; social breakdown; psychological trauma.

Indirect Disaster Effects

Long term recovery, migrants and refugees, disruption of economy and supply chains; famine, civil unrest, armed conflicts

Emotional Impacts

Negative effects on your mood, outlook, and welbeing even when you are not physically or materially in harm's way

Three Classes of Mental Health Impacts of Climate Change
Source: Adapted from Thomas J. Doherty and Susan Clayton, "The Psychological Impacts of Global Climate Change," *American Psychologist* 66, no. 4 (2011): 265–276.

As climate disruptions grow in frequency and severity and increasingly converge, these direct, indirect, and emotional impacts often occur simultaneously, causing us to experience the scale vertigo we discussed in Chapter 1. Global becomes local. What you thought of as distant problems suddenly manifest right at home in the form of choking smoke, fire evacuation orders, or a flooded home.

For Reid, looking at the direct, indirect, and emotional impacts of the climate threats to the Gulf Coast region provided clarity. He was directly impacted by the hurricanes, storm surges, flooding, and evacuations he'd experienced. He also faced a number of indirect impacts, including a chronic cycle of post-disaster recovery and cleanup. Adaptation does not always mean rebuilding in the same place; sometimes it means an opportunity for finding new directions. Reid had personal doubts about rebuilding in a disaster-prone area near a huge oil and refining infrastructure and did not expect Texas to wean itself from fossil fuels in the near future. Reid was also more aware of the struggles of his community as seen through the eyes of his high school students, many of whom lived in low-income neighborhoods that were often the most damaged and least protected.

Reid and Marisol had long harbored a dream of moving to the south end of the Gulf Coast on South Padre Island, close to the border with Mexico, where they and their children could fish, surf, and witness the sea turtle migration. As a teacher and nurse, Reid and Marisol could work in the underserved south Texas colonia areas, populated largely by impoverished Latino and immigrant communities that have grown up over the years in the borderlands. But after a major storm and tornado hit nearby Port Isabel in May 2023, and with more storms on the horizon, the couple reluctantly agreed to let go of their seaside dreams.

Now they were considering a move north to Dallas, chosen for the well-regarded school system and distance from the perilous coast. But Reid was torn about this decision. While it was best for the young family, he felt like he was abandoning his own parents, siblings, and grandparents. He felt uncomfortable with the privilege of being able to relocate and experienced survivor guilt. Paradoxically, the ability to adapt can also make us feel guilty for our privilege. But that is more a fault of the system than of us.

Reid and Marisol knew that they needed to relocate and wanted to pick a place where their family would be safe. Was anywhere safe? A tool they found helpful was the Climate Vulnerability Index, developed by the Environmental Defense Fund and Texas A&M University, which integrates climate change impacts with other environmental, health, and socioeconomic factors, such as living in proximity to fossil fuel refineries and other polluting industries.[3] This gave Marisol and Reid an objective picture of their risk situation and helped them make decisions that they felt were critical to the safety and mental well-being of the family.

The U.S. Climate Vulnerability Index

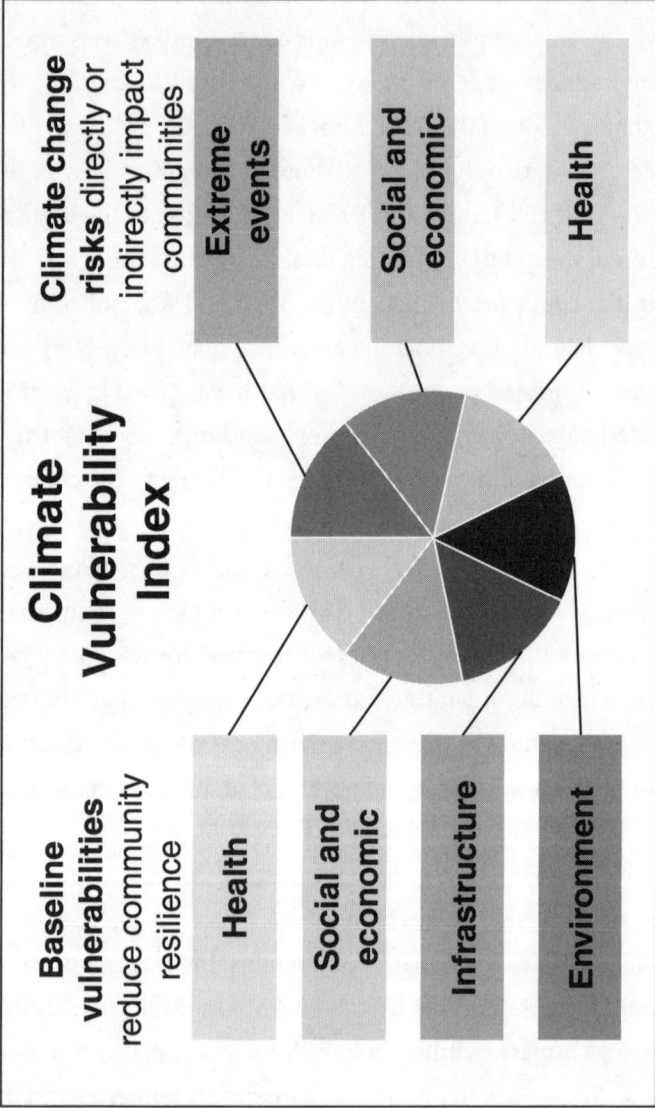

Baseline vulnerabilities reduce community resilience

Health

Social and economic

Infrastructure

Environment

Climate Vulnerability Index

Climate change risks directly or indirectly impact communities

Extreme events

Social and economic

Health

Climate Vulnerability Index
Source: Climate Vulnerability Index, https://climatevulnerabilityindex.org.

CREATING YOUR OWN PERSONAL SUSTAINABILITY REPORT

Your personal IPCC assessment has four questions to ask yourself to help clarify the most serious mental health impacts that affect you. We know climate breakdown is a capital-*I* issue, but what are your unique little-*i* issues? The first step is thinking globally but assessing your impacts locally.

1. *How climate breakdown affects your local weather.* How are ongoing changes to the global climate affecting your *local* weather, the *local* natural environment, and the community where you live? To answer this one, you'll need to explore changes and trends in your region and your hometown. In the United States, the recent National Climate Assessment has traditionally been a great resource that reports impacts on the economy and people's health, with an interactive atlas that zooms in to the region, state, and county levels.[4] I'll also share more resources at the end of the book.

Depending on your region, you need to objectively consider several climate-fueled risks. Just as you worked through your emotions earlier, give yourself space to reflect on them. Notice which are more salient for you:

- Extreme and extended periods of daily and nighttime heat
- Drought and water shortages
- Wildfire, especially for those living at the wildland-urban interface
- Unhealthy air quality and smoke
- Extreme rainfall and flooding
- Hurricanes, cyclones, typhoons, and nor'easters
- Tornadoes and high winds
- Severe blizzards and snow and ice storms

2. *Your mental health impacts.* Now, look back at your life. How have the three major categories of climate impacts—direct,

indirect, and emotional—affected you in the past? What kind of impact, or combination of impacts, is most pressing, doing the most harm, most impacting your quality of life, or causing you the most anxiety right now? What do local weather trends portend for future impacts?

3. *Your buffers and risk factors.*

    (a) I call the positive supports you have *buffers* since they cushion the blow from disasters and help you to be resilient. What local factors and personal resources help protect you from mental health impacts? This could include a sheltered location, your personal skills, social support from your family or community, and economic advantages (such as insurance, ready travel or evacuation options, and ability to rebuild or relocate).

    (b) What are the *risk factors* that place you or your loved ones in harm's way? This could be your location if it's vulnerable, as well as limited social or family support, and financial precarity (insecurity about your employment or income). You are more vulnerable if you are very old or very young, if you have a preexisting health condition (such as one that needs regular medical care or medication), and if evacuation is difficult or impossible for you. Now, take a second look. Think about other social factors or *na-tech* disaster threats (threats of a technological disaster triggered by a natural hazard) that compound climate risks in your area:

    - Social, ethnic, class, or other conflicts that impede community functioning
    - Living in an earthquake zone
    - Proximity to nuclear or chemical leaks, spills, or explosions
    - Wars and regional conflicts

3. *Lenses, blinders, and blind spots.* This is where you practice some metacognition—thinking about your thinking—about how you

are assessing your mental health risks, and how this affects your emotions and stress level.

(a) A *lens* helps you to see potential impacts clearly, providing either a big-picture view or a hyperlocal view. This is the place to consider the sources and media through which you learn about climate and environmental impacts, and the quality of this information. Is what you are accessing helping you to see clearly and accurately, and providing concrete opportunities for action (such as the solution-focused journalism we looked at in Chapter 3)? Is it balanced, offering an overarching view of climate trends and risks and information that is directly relevant to my location (like the US National Climate Assessment)?

(b) For the purposes of climate coping and adaptation, a *blinder* is something that helps you filter out distracting information, much like a horse blinder helps the animal move ahead without being startled from the sides. A well-designed adaptation plan helps you to focus, apply your energy efficiently, and not get caught up in issues that are not crucial for you.

(c) A *blind spot* is an area of missing information, either a risk factor you are not aware of or a buffer you are not appreciating. Research shows that people selectively scan for information based on their preexisting beliefs and their moods. Even daily weather can affect how we interpret information about climate issues. Are you getting distracted by unimportant information, or discounting or ignoring important information based on your untested assumptions or mindset? Can you check for blind spots by seeking others' feedback?

Identifying your blind spots can help you detect hidden threats as well as the ways you might be *underestimating the resources at your disposal to buffer you against those threats.* For example, are you aware of programs that work to limit the climate change threats that affect you or that help you or your family to adapt?

All Environmental Issues Are Justice Issues

Once you consider the role of resources and buffers in your own life, you realize who has them and who lacks them—and the massive inequity involved. All around the world, marginalized communities who have done the least to contribute to climate change bear the highest costs. One stark example is the large number of marginalized people living in *sacrifice zones*: places damaged or poisoned through destructive land use, or where residents are forced to live adjacent to heavily polluting industries. Not only are sacrifice zones profoundly unethical and a violation of human rights, but they are also steeped in racism. As Hop Hopkins from the Sierra Club famously stated in 2020: "You can't have climate change without sacrifice zones, and you can't have sacrifice zones without disposable people, and you can't have disposable people without racism."[5]

Because Larysa's eco-anxiety seemed so free-floating, the concrete framework of her IPCC assessment helped her to apply her project management skills to work through basic questions about the direct, indirect, and emotional impacts that most affected her as a postpartum mom. Consider how she approached these questions as you begin to formulate answers of your own.

Question 1 required her to step back and consider *how global climate changes and disruptions were affecting the cycles of her local weather*. In the Pacific Northwest, where Larysa and I live, warmer winters bring less snowpack, leading to droughts, while hotter days and temperatures that don't cool down at night bring dangerous heat waves and increased frequency and severity of wildfires, with lasting smoke and air quality issues. Meanwhile, the warmer winters bring damaging extreme rain and erratic snow or ice storms.

Question 2 helped Larysa see that scary experiences with ice storms, power outages, and extreme heat waves had had the *most urgent mental health impacts* on her. Though they did not necessarily reach the level of community-wide disaster, they overwhelmed her ability to cope and respond.

When it came to question 3, about her *personal buffers and risks*, Larysa was fortunate. She recognized she was relatively buffered by social and economic resources, and that her and Declan's livelihood and home were secure.

Question 4 helped Larysa see how her news feed and prepper research had become a problematic *lens* that exaggerated her sense of immediate danger and left her *blind* to more positive stories. Remember that we develop our knowledge about climate and environmental issues through many channels: our personal experience, messages from our family or social groups, news, social media, and education. The capital-*I* and little-*i* aspects of the issues help determine what information stands out for you, but they can also leave you vulnerable to *blind spots*.

TIME HORIZONS

Once we identified Larysa's primary threats, risks, and buffers, I gave her a set of *time horizons* to help her plan her next steps: the next six months, six months to two years, and two to five years. You can place items in these time boxes to prioritize what needs to be acted on first.

The time horizon of the *next six months* is the most urgent. This could include actions such as recovering from a disaster or preparing a go bag for evacuations.

The horizon of *six months to two years* (which could include something likely to occur over the next hurricane or fire season) is the realm of *disaster preparedness* and actions that take time or investment, like saving up to buy a backup battery system with portable solar panels, learning about local disaster response and citizens groups in your area, talking with family about evacuation needs (including extended family and relatives who are at higher risk, like people with disabilities, elders, or those with young children), and considering how climate and weather events will impact your job and business.

A *two-to-five-year* time horizon is more of a long-range plan, and can include choices you make about your life direction, including your education, relationships, retirement, or major changes like relocation or a new career direction. For example, Reid and Marisol's risk assessment led them to make changes they felt were critical to the safety and mental and physical well-being of their family. One change was relocating. They realized that a two-to-five-year time frame was not soon enough. Immediately Marisol applied for nursing jobs in Dallas. They began looking at housing and schools. Within two years they wanted to have completed their move.

Rachel

Sometimes your IPCC focuses more on your family or others rather than yourself.

Rachel was a mental health therapist in one of my training groups. She lived in a big city on the East Coast, though her denim shirts and turquoise jewelry tipped me off to her roots in the Southwest. She'd spent her childhood as a "free-range kid" in Boulder, Colorado, her self-described hippie parents allowing her to disappear into the woods to play until sunset. Rachel typically considered herself a relaxed parent, encouraging her own children to explore the world and take risks. But now she struggled with a little-*i* concern she was keeping to herself.

Rachel's eldest daughter was set on attending a prestigious college in Southern California. Rachel confessed she was extremely worried about wildfire risk given high-profile fires in nearby Malibu. The thought of having her daughter living so far from home, and in such proximity to those dangers, led her to feel anxious and lose sleep. But she was not sure how to broach the subject with her husband and daughter for fear of being seen as alarmist, irrational, or overprotective. Her therapist friends just attributed her worry to fear of an empty nest. I took Rachel's concerns seriously. California is a region prone to wildfire, and given the warming climate, her worries about the direct impacts on her daughter from severe fires were quite rational and very practical.

Rachel's IPCC assessment on behalf of her daughter lined up the local weather changes, potential disaster impacts, and her daughter's risk. However, question 4 soon revealed some blind spots. Rachel's anxiety was based on reading news about wildfires, without knowing what the school her daughter wanted to attend was doing to protect students from the danger. I knew from my own university teaching and involvement in organizations like the

Association for the Advancement of Sustainability in Higher Education that colleges were taking disaster planning seriously, and so I encouraged Rachel to research the college's disaster planning. When she did, she discovered that all the Southern California colleges her daughter was looking at addressed wildfire risk in their disaster planning. Some schools were even nationally recognized leaders in wildfire response. Rachel was able to share this information with her family and in the process normalized conversations about climate risks and disaster response as part of her daughter's college planning.

Rachel's experience illustrates some larger points about climate psychology and how we cope with threats. A common approach to education about climate change (known as the *information deficit model*) attributes barriers to coping or action to lack of information. The assumption is that if experts can just teach the public about an issue (such as links between fossil fuel use and climate disasters), then people will see its importance and take actions accordingly (such as using less fossil fuel). This is a perfectly rational approach. And as with Rachel, the information deficit model is an important tool that reminds you to look first for your knowledge gaps. However, in many situations information alone is not enough to spark changes in environmental behaviors. Later in the book we'll confront the fact that there are often competing issues and priorities, or political, social, or economic barriers that can get in the way of taking a desired action.

MENTAL SKILLS FOR SUSTAINABLE COPING

Your IPCC report will ultimately help tame your anxiety by dispelling mystery (a core feature of anxiety) and giving you clear and actionable ways to address the threats that climate breakdown poses for you, your family, and the animals and places you care about. However, distilling anxiety into clear risks and realistic fears calls on your coping resources and sometimes requires building new capacities, using skills we discussed in Chapters 1–3.

1. Approach your IPCC with a *growth mindset*. It will make you more likely to be open-minded and feel curious and creative. If you

approach this with a fixed mindset that focuses on your feelings of doubt, apprehension, and helplessness, then yours becomes a grim, scary task that you would more likely avoid. Allow yourself the possibility of learning and changing for the better.

2. Recognize how you can *assimilate* information about climate and weather trends to what you already know about the weather in your region. Make room to *accommodate* new information about *unnatural disasters* that previously would have been unheard of for your region (like wildfire in typically damp forests, or catastrophic flooding far inland from coastal storms). Being able to take in new information and adapt is a sign of *psychological flexibility*, and that's key to sustainably coping.

3. Be honest and expressive about your emotions. If you can, do your IPCC research with a supportive group where you can share and be validated. Keep in mind what you *want to feel*. I suggest adding patience, curiosity, determination, and taking pride in making the effort.

4. Soothe your body in the present moment by keeping your breath open and taking time outdoors to get a reality check about the place and weather as you find it outside your door today. As we have learned, a key part of coping with runaway climate anxiety is the ability to access a sense of *present-moment safety*.

DEEP ADAPTATION

Like many of the big ideas and concepts we grapple with in this book—nature, sustainability, identity, wilderness, happiness, duty, and action—climate adaptation can be approached from different perspectives, shallow or deep. Conventional approaches to adaptation and related ideas like *resilience* often assume that our existing systems and lifestyles can be preserved as they are. However, there are other options. *Transformational resilience* refers to the opportunity to break with unhealthy aspects of the status quo and create new ways to live that support the rights of humans and other species.[6] Another perspective is that adaptation, as it is typically discussed, is simply not possible. The effects of climate breakdown are so serious and baked into global systems that learning to prepare for the possibility of societal collapse is a wiser, even necessary, option.[7] We'll consider these ideas more as we get into identity, values, and action. The question for you to consider now is what you want to do with the *opportunity* that your adaptation can afford.

ADAPTATION PRECEDES MITIGATION

The traditional approach of climate science and policy is logical and systematic: First, understand the problems and create a rationale for *climate mitigation* (lessening the severity of climate and weather damages) by slowing, stopping, and reversing the underlying process of global warming (for example, by transitioning to safer and cleaner energy sources or using carbon removal technologies). Then focus on *climate adaptation*—learning to accept and to live in an already changed and diminished or more dangerous world. That makes a certain sense. Isn't it better to remove a problem rather than learning to live with it? From this perspective, starting with adaptation could be seen as a form of giving up or giving in.

This mitigation-first approach was appropriate decades ago when the possibility of getting ahead of global warming was achievable. Now major damages from climate change have already occurred, and even with all the new mitigation options on the scene, we have a pressing need to learn to survive and thrive in the new world we have.

I find the approach in which *adaptation precedes mitigation* to be more in line with how people actually deal with setbacks and disasters. It's hard to muster the motivation to address an abstract, "maybe" threat. Once we actually live through and learn the true dangers and costs of storms, floods, and wildfires, however, we are much more motivated to mitigate them for the future. The process of adapting is not giving in. It is building your capacity to navigate these threats with confidence, so you can begin working proactively toward mitigating climate breakdown for the future.

SOMETIMES IT'S BETTER THAN YOU THINK

As we've discussed, the more you look into climate science, the more you might find that the problems are more severe, the stakes higher, and barriers more vexing than you think. But when you look deeply, sometimes you find that it's actually *better* than you think. There are many good things happening, thanks to the efforts of people all around the world. So you have to build capacity in two ways, to take in both the bad *and* the good.

Yes, disasters are getting worse (larger, faster, more frequent, more expensive) and disaster deaths are on the rise. But so is our collective ability to predict when disasters will strike and respond to and cope with the dangers, thanks to wide-ranging communications, more sensitive early warning systems, more resilient buildings and codes, and increasingly sophisticated emergency preparation and responding.

Millions of people go to work every day in government agencies, businesses, schools, nonprofits, and grassroots organizations whose mission is disaster preparedness and response. As Larysa and Rachel discovered, one of the best ways to educate yourself about local disaster concerns is to find out what agencies in your community are already doing to protect you.

YOU ALREADY KNOW

Climate change is the greatest public health threat in history precisely because it makes almost all existing public health problems worse: addiction, asthma, cardiac disease, child abuse, crime, depression, drought, family problems, famine, homelessness, infectious diseases, life span, mental illness, refugees, suicide, violence, war. All of these are negatively affected by global climate change. This illustrates the hyperobject nature of the problem. It's all around us, and its elements are related in ways you might not expect. But when you understand that climate change is an interconnected *ecological* problem, it makes sense there are many mechanisms and pathways linking climate disruptions and health outcomes.

Once she worked through her IPCC, Larysa found that each new climate story no longer activated her fight-or-flight response or knocked her off her footing. She had a better understanding of global threats, her buffers, and risks—and, best of all, she no longer felt the need to chronically doomscroll. And when she did see a negative story, she would scroll right past it, saying to herself, "I already know."

"I already know." These three words are enormously freeing. Begin to trust your understanding and your knowledge. When you hear a new public health story about climate dangers, you can say, "I already know." Because you do. As someone who has been looking at climate science for many years, I find that most dire stories and predictions tell me things that I already know and have known for years. It's the positive news that is usually new for me.

The theme of Part One has been "Open mind, open heart, open breath, open hands." You've opened your mind to how you think, your heart to how you feel, your breath to be more present, and your hands to grasp coping skills for balance. Next, in Part Two, I'm going to give you a whole new way of thinking about yourself in relation to nature, and a superpower you can use to combat environmental despair.

PART TWO

Identity

IN THE NEXT FOUR CHAPTERS, I INVITE YOU to explore an identity as empowering as gender and ethnicity: your environmental identity. I'll show you how to look back on the eco-story of your life and identify your unique nature values in order to understand why the environmental crisis can hurt on such a deeply personal level. What we inherently value is tied to our identities. To deepen and expand your environmental identity, I'll give you a framework to understand your relationship with natural spaces, starting with your own home and the local landscape. We'll also explore how your environmental identity and values manifest in your family history and intimate relationships, and how relationships with nature can span a lifetime, from infancy to elderhood. With this knowledge, you'll be more prepared to take on the self-healing skills needed for more serious issues like eco-anxiety and despair. You'll begin to see that we hurt where we care. We care where we love. And where we love is where we find ourselves.

Chapter 5

Meaning

I have established a long-term partnership with the force on Earth that wants to live...

In the process, I have changed drastically. My identity, priorities, my spiritual beliefs, what makes me feel good about myself, and how I spend my time.

—Margaret Klein Salamon[1]

It is the year 1998, on the eve of the millennium. A group of environmentalists and conservation professionals have gathered in southern Vermont to take part in a retreat led by the Buddhist teacher and environmental philosopher Joanna Macy. For years, Macy has been convening groups around the world to help people find emotional support and community as they struggle with concerns about issues like nuclear war and global environmental decline. Her approach, called *despair and empowerment work*, features group-based exercises and rituals that allow people to begin to let down their defenses and express their love, fear, anger, and grief about nature and the natural world.[2] Honoring these feelings leads to an experience of emotional catharsis, a liberating of pent-up energy, and potential for even deeper emotions and connections.

The culture of these workshops is very hands-on, inviting participants to be physically close, look into each other's eyes, and openly share their emotions. One activity from the workshop that day, the Bodhisattva Check-In, was inspired by the Buddhist ideal of a person who vows to seek enlightenment and serve other beings.[3] Participants were asked to imagine how the unique details and conditions of their life (such as where and when they were born, or their skills or abilities) could serve the healing of the world—almost as if they had chosen their life circumstances for this purpose.

Imagine you are one of the participants at the workshop, seated in a circle on the floor of a large barn-like meeting space. Some soft classical music plays. You're invited to close your eyes. You listen as Joanna recounts the vast history of planet Earth, its billions of years of time, the great geological and biological eras, the emergence of life, and the branching of evolutionary history to the present day. Joanna explains the Buddhist concept of bodhisattvas, enlightened beings who vow to keep returning to the world to relieve suffering. She asks you to imagine that you too are a bodhisattva and can remember how you *chose* to be alive as a human in your particular moment of history.

You and the others in the group imagine being together in the "time out of time" before your births. Joanna then tells of a crisis facing Earth at the end of the second millennium:

> The challenges take many forms—the making and using of nuclear weapons, industrial technologies that poison and waste whole ecosystems, billions of people sinking into poverty—but one thing is clear. A quantum leap in consciousness is required if life is to prevail on Earth. Hearing this, we decide to renew our commitment to life (our bodhisattva vow) and reenter the fray—to birth as humans in the twentieth century [the year was 1998], bringing everything we've ever learned about courage and community.

You reflect on being born in such a challenging time. When you are ready, you and the other participants stand up one by one, symbolically "choosing" to be born. A drum beats as people begin to circulate around the room. Joanna asks you to look around at the faces of your fellow bodhisattvas and honor their choices to return, and then to find someone, wordlessly connect with them, and sit down together.

Then the main part of the exercise begins. You and your partner take turns telling each other about the particular life you have assumed, and the situations you chose to inhabit. Joanna guides with questions (and consider these for yourself as you read): When and where were you born, including the timing and season? What was the place and landscape you chose? What social conditions? What religion or faith tradition, or lack thereof? What gender? What parents or caregivers, siblings, or birth order did you choose? What unique abilities or disabilities did you choose to accept, and how do they help you connect with the planet and those you may serve? What mental, physical, or spiritual desires did you summon for yourself? How might you, at this point in your life, understand what your mission could be in terms of serving Earth or making the world a better place?

Exercises like the Bodhisattva Check-In can be a helpful and empowering thought experiment. It's also a view into a spiritually attuned version of environmentalism that is meaningful for many people.

Among the group that day was a graduate student studying psychology who had come to learn about Macy's techniques, while also exploring his own environmental consciousness. I know this young man quite well because that person was me. As it happened, we had an odd number of people in the group and I did not have a partner for the sharing exercise. So I got to recount my birth choice story under the compassionate gaze of Joanna Macy herself.

The insight that I shared with Joanna was just how pivotal the time of my birth—June 1965—seems. Seeds of so many cultural movements that define the early twenty-first century were being planted. It was the height of the civil rights movement in the United States, with the Selma-to-Montgomery march and the passage of the Voting Rights Act occurring that year. The day after I was born, astronaut Ed White became the first American to walk in space. The year 1965 saw the first uses of the terms "globalization" and "sustainable development." The Wilderness Act, one of the first laws in the United States recognizing the inherent rights of nature, was passed, as were the Water Quality Act, the first automobile emissions standards, and the first regulations requiring health warning labels for cigarettes. I've also learned more recently that in 1965, US president Lyndon Johnson received a prescient briefing about global warming from leading atmospheric scientists of the time.

I used to say that it was accidental that I got into climate change work. But when I think about the historical moment in which I arrived, it seems almost inevitable that I would choose this path. The Bodhisattva Check-In stays with me because it was a milestone in the development of my environmental identity. What might you find if you considered *your* birth year from this perspective? How might it feel for you to imagine "choosing" to be born in your particular time and place?

ENVIRONMENTAL IDENTITY

Each of us has a complex evolving personal identity that has many facets: gender, sexuality, politics, ethnicity, culture, race, religion, and many more interconnected identities. We also have an *environmental identity* that few of us have actively explored.

This aspect of our identity, studied by conservation psychologists, refers to our deep-seated beliefs, attitudes, and relationships toward nature in all its forms. This includes your self-image, your underlying values and beliefs, and how "connected" you feel with nature. Your environmental identity is shaped by many events and forces, like your family, your culture, and the place and time you were born. Understanding your environmental identity provides a stabilizing, grounding force to combat the free-floating existential dread you may feel about the planet's future, and a crucial ingredient for health, meaning, and happiness in the modern world.

Jann
Jann, a conservation biologist, had worked for years as an administrator for the Canadian National Parks. Her role involved trying to balance the needs of large free-roaming animals, like grizzlies, with those of a public

eager for outdoor recreation. While Jann was outwardly successful, inside she felt hopeless and adrift. Her unique window into ecological issues such as threatened species, wildfires, and disappearing Rocky Mountain glaciers created a complex and sometimes overwhelming sense of loss. She came to see me because she was feeling exhausted and burned out. To recover her purpose, Jann needed to first gain insight into her environmental identity.

I introduced Jann to an *eco-timeline* exercise that helped her identify key influences, events, and turning points in her life connected to nature and the natural world. We all have milestones and experiences in nature that mold our environmental identities over time, from where we grew up to formative memories and experiences in the present day. Because there is value in the hands-on exercise of actually sketching this, I had Jann draw her eco-timeline, using a large piece of paper and colored markers. First, she drew a line representing the span of her life from her birth to the present. Then I invited her to search her memories, identify significant experiences in nature, and place them on the timeline. These included the region and environment where she grew up, her favorite places, family activities in the outdoors, contact with other species, travel, teachers and mentors, influential books and movies, consciousness of larger world events, and losses and injuries related to nature and the outdoors. As she sketched, I suggested other experiences to consider, such as her education, work, and parenting.

I find that most people quickly get immersed in this exercise, and Jann was no exception. As poet and ecological thinker Gary Snyder has noted, we carry within us a picture of the terrain we learned as children, and it can be remarkably easy to conjure sights, smells, and memories.

By laying out her own eco-timeline, Jann could see how her adult relationship with nature was shaped by her childhood activities with her family in the forests and lakes of Ontario, books she'd read like *Julie of the Wolves*, the time she spent in her biology field work closely observing landscapes, and her awakening to global environmental issues gained through her travels as a young adult. She saw how her idealism had waned as she climbed the career ladder in her male-dominated field and experienced setback after setback in her goal of preserving wild places. Now her children were grown and starting their own families under the shadow of climate change.

This exercise provided a window into Jann's environmental identity as a protector of wildlife and nature. Like other important aspects of identity, our environmental identity is not always self-evident. We might not even be aware of it until it is brought to our attention. Like other facets of ourselves, it needs to be recognized and given language so it can be expressed.

So far in the book, you've explored how you can mentally adapt to a climate-changed world by being more aware of your mindset, your feelings, your stress management, and your unique environmental threats. Your environmental identity provides a healthy container for all these and more. As you will see in the next few chapters, your environmental identity relates to your core values and experiences in the natural world. And these in turn influence where you find meaning and purpose along your life path. In later chapters, being familiar and secure with your environmental identity will assist your ability to cope, heal, and take action. One way to get in touch with your environmental identity is to create your own eco-timeline.

YOUR ECO-TIMELINE

Now it's your turn to draw your own timeline. If possible, sketch this out on paper using words, pictures, symbols, or some combination of the three. Draw a line representing the span of your life from birth to the present, leaving some open space on the end to fill in later. What memories or events come to mind? Perhaps you had an outdoor retreat that was really important to you, or a special relationship with an animal or a pet. Perhaps you learned something memorable about the environment in school, or family or friends taught you about camping or being outdoors. Maybe you had negative or scary experiences outdoors, or were taught that being in nature was dirty or unsafe. Everyone is different. Charting your experiences on a timeline is a great way to get in touch with the unique way you think of your own personal identity in relation to the rest of the natural world.

When you begin to populate your timeline, you may find it easy to begin with firsts, such as the first places in nature you remember visiting, your first childhood pet, or early memories of outdoor activities such as playing outside, farming, camping, gardening, hunting, fishing, hiking, or outdoor sports. It's natural to gravitate to positive memories and experiences, identifying positive events and emotions in your timeline will help you see more of these going forward.

But it's also important to include uncomfortable or painful experiences: injuries, fears, losses, disasters, and traumas (a time you became lost and afraid as a child, or you witnessed an accident, or found yourself a victim of injury or harm). Identifying both positive and negative experiences in your timeline from the beginning will help you see patterns and trends in the story of your life. It is always amazing to me how even distant memories come to the surface in this exercise. Your environmental identity is just under the surface, waiting to be recognized.

Elements Your Timeline Might Include

- The landscape or climate of the region where you were born or spent your childhood
- The first place you remember playing outside and experiencing being in nature
- First relationships with a pet or wild animal
- How you interacted with or spoke of nature with your family, siblings, or friends
- Injuries and scares (accidents, getting hurt, being fearful of deep water, ocean tides, or great heights)
- Feeling uncomfortable, out of place, excluded, or threatened when spending time outdoors or doing nature-based activities
- First time camping or sleeping outdoors (including summer camps)
- First time gathering, foraging, gardening, farming, hunting, or fishing
- Contacts with "the wild" (places, animals, weather)

- First encounters with special landforms—forest, desert, mountains, rivers, oceans
- First time seeing stars or the Milky Way
- First experience of exploring nature on your own
- Outdoor skills you became good at (like identifying species of birds, growing a flower garden, sailing a boat, or navigating in the woods or mountains)
- Travel and vacations where you experienced different topographies or climates, or spent a lot of time outdoors
- Key role models, teachers, or mentors from whom you learned about nature (teachers, scout leaders, elders, siblings, or friends)
- Participating in faith or spiritual traditions connected to nature
- Peak outdoor experiences (great joy, mystical experiences, epiphanies, or "waking-up" moments)
- Tragedies, trauma, or loss (of place, people, or other species)
- Rites of passage (formal or informal) in or related to the natural world
- Key books, movies, music, or art related to the natural world
- Introducing a child to nature or to the outdoors as a parent or grandparent
- Jobs and professional roles that put you in touch (intellectually or physically) with the natural world
- Participating in community organizations, politics, or activism with an environmental or conservationist focus
- Embracing the role of environmental advocate or teacher, and passing wisdom and care to future generations, or creating lasting projects
- Things left to do: life goals for travel, adventure, conservation, and learning about nature

TELLING YOUR ECO-STORY

At the heart of your environmental identity is your *eco-story*, the narrative of your relationship with the natural world and how you've been touched by environmental issues throughout your life.

My own eco-story is one that travels a wide span both geographically and emotionally. As a boy, my idea of nature was limited to the life I knew in 1970s rust-belt Buffalo. If you had asked me to point out "nature," I would have referenced the grass on my lawn, birds and trees in the park, the woods where the men and boys would go fishing up north in Canada, and images of exotic places I saw in nature documentaries. I wouldn't have considered *myself* to be a part of nature, nor thought of myself as an environmentalist. For me the first Greenpeace voyages, the first Earth Day, and other seminal events of the late 1960s and 1970s are essential history lessons. But I came of age in the 1980s, in the reactionary Reagan-Thatcher era, not exactly a time of peace, love, and understanding. I'm more punk rock than Woodstock; the soundtrack for my adolescence was the Sex Pistol's "God Save the Queen," with its line "No future for you."

But as I grew to adulthood, a series of incredible new chapters of my eco-story were written (and I actually believe in a future much more now than I did growing up). It started in the biggest of cities. I went off to college in New York City and experienced life in a metropolis, less nature and more concrete, except for special places like Central Park or the ocean at Rockaway, not far to reach. I also spent time studying in the west of Ireland, which in the late 1980s was still a radically slower, quieter, and more pastoral country where I could roam about the farms, green misty fields, stone fences, and grounds of ancient castles.

After college, my first jobs were in some of North America's wildest, most remote places. I hitchhiked to Alaska and found a job on Kodiak Island on a salmon fishing boat. There I got to see—and hear—a true salmon run, thousands of migrating salmon returning from the ocean to their home rivers, a thunderous force of nature. I worked at the isolated North Rim of the Grand Canyon and as a rafting guide deep in the Canyon on the Colorado

River. I helped lead three-week-long backpacking treks in the high desert in outdoor therapy programs for teenagers. I became someone who was out-doorsy and adventure-seeking. As a child, anything I knew about the vast landscape of the American West I would have learned from watching shows like *Grizzly Adams* on TV. But now I was immersing myself in places, in US and Native American history, and began to see the landscape as alive. I became aware of the natural environment as more than just a backdrop or an image. While working for Greenpeace, I had an extended multiyear experience of waking-up syndrome.

I also began to experience firsthand all the positive and healthy things that exposure to nature did for me, for struggling young people, and for the well-heeled tourists I guided. I couldn't yet see the big picture of my timeline, but my experiences were leading to my discovery of *ecopsychology* and the connection between mental health and healthy nature. It was all a precursor to my training as a clinical psychologist and my subsequent engagement with issues like eco-anxiety and climate change. And, like you, I keep evolving and extending my eco-timeline today.

INFLUENCES ON YOUR ENVIRONMENTAL IDENTITY

The Relation Between Your Self and Nature

Years ago I had the opportunity to create conservation psychology training programs with the very people who developed the field, including Wesley Schultz and the late Carol Saunders. One basic way that psychologists like Wes study environmental identity is by measuring people's "inclusion of self in nature." The illustration below presents pairs of circles representing your-self and nature. The circles in the first drawing suggest your relationship with nature is separate, and each successive drawing shows more and more overlap between the two circles, portraying yourself and nature as increas-ingly connected. The final drawing, number seven, illustrates a sense of one-ness or interbeing.

Simple Assessment of Environmental Identity

Self & Nature

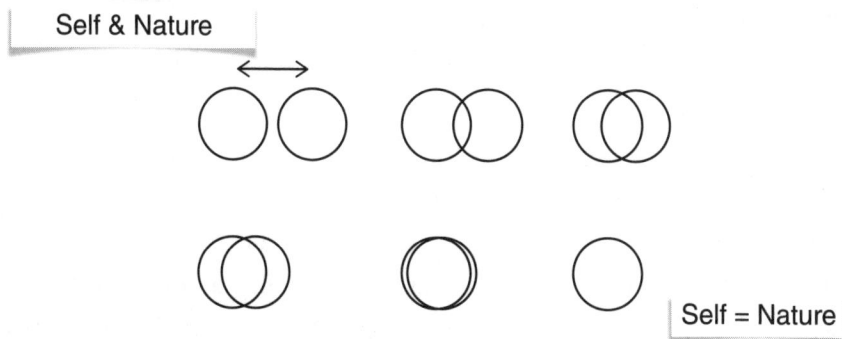

Self = Nature

The Inclusion of Nature in Self Scale
Adapted from: P. Wesley Schultz, "Inclusion with Nature: The Psychology of Human-Nature Relationships," in *Psychology of Sustainable Development*, ed. Peter Schmuck and P. Wesley Schultz (Boston: Kluwer Academic, 2002).

Larysa, the nursing mom we met earlier, valued nature but typically considered herself separate from it. Marcus, the Portland city planner, recognized a clear relationship between nature and his health, so he chose circles that slightly overlapped. Glacier researcher Roman and wildlife biologist Jann recognized a deep interconnection with nature, so their circles were strongly overlapping. Delphina, the veterinary student, and Reid, the teacher and surfer from hurricane country, decided their self and nature circles were one and the same.

There is no "right answer" to this self-test. While these kinds of surveys help measure our baseline beliefs, these perceptions are dynamic and constantly shifting depending on our context. If I am stuck in a rush-hour traffic jam or lying on my back getting an MRI scan in the hospital, I might feel *very* separated from nature. I might even feel separate from my own body. But if I am digging in my garden, walking down a tree-lined street on a sunny spring morning, or watching a sunrise or a sunset, I might feel highly connected with nature. If I am feeling a deep, meditative sense of flow while paddling in the surf, having a psychedelic experience, or staring up at the Milky Way on a clear desert night, my self and nature circles may be one.

Diversity and Intersectionality

The concept of *intersectionality* describes how different aspects of your identities overlap to create your unique life experiences, privileges, and vulnerabilities. This is a helpful lens through which to think about your eco-story. Look again at your timeline. Do you think that your experience would have been different if your skin color was different, or your gender, or your physical ability, or your sexual orientation? Or if you grew up in a different social class, or a different place in the world? Even within families people can have very different experiences of nature. For example, Canadian biologist Jann had different experiences of the outdoors than her brothers, who were encouraged to play hockey and go hunting, while she was encouraged to garden, tend to injured animals, and develop her artistic side by painting landscapes.[4] Now you can get a sense of the great diversity at play in people's experiences of nature—what I call *environmental diversity*. No matter where you live, environmental problems have the most impact on those with the least amount of power and privilege; one's experience of nature is inextricably linked to questions of discrimination, justice, and freedom.

You can do the eco-timeline exercise alone or as a group activity. I've done this with people from around the world of varying backgrounds and cultures. If you do it with others in a group, it's a great opportunity to be curious and humble, and to be open to different ways of seeing the world (or the climate elephant). Even though I try to be conscious of multicultural diversity and factors like privilege or the presence of harm, I always learn something new.

Environmental Identity and Life Meaning

Each of us has a life path and a sense of meaning. Some have a very explicit and meticulous plan; perhaps they are an athlete or artist with specific performance goals, or know exactly what career they want to pursue, or have a vision of the family they want to create. Moreover, aging and illness can impose a sense of urgency that makes a life path clear. But for many of us, the idea of a life path and meaning isn't something we think about every day. I like to think that each of us has *two paths* in our life, the actual life that we live, and the aspirational idea of what our life should be that's operating under the surface. Like a mole traveling under the ground following its keen sense of smell, every once in a while we pop up and take a look to see how we are doing on our path.

One way to understand the mental challenge of climate change is that it has crossed your life path, bringing your unconscious and unexamined environmental identity into awareness. The world that you think you should have, or have taken for granted, is now threatened and insecure. So in this view, *the crisis of climate change is a crisis of personal meaning.*

There are probably as many ways to think about meaning in life as there are people. For our purposes, I recommend you begin with a three-dimensional approach to life meaning:

- *Purpose.* Do you have a role or something you are striving for?
- *Coherence.* Does the life you have make sense to you given your goals and your background?
- *Significance.* Do you have a sense your life is worth living or that it has value to you or to others, that you are making a positive impact?

Getting to know your environmental identity is a helpful new lens through which to look at life's meaning—something that is essential in an era of climate disruption. (Now you can also add environmental identity to the lenses you can look through in your IPCC to help direct your adaptation efforts.)

As Jann looked over her eco-timeline, suddenly she was able to see the whole of her life as a coherent story leading to the present moment, from her childhood connections with nature and her youthful dreams, to her choice to devote her career to science and conservation. For a large period of her life, she had a purpose as a wilderness manager in Parks Canada. She felt privileged to have found so much meaning in her job. Using her skills as a biologist and administrator, she helped preserve a whole web of life from grizzlies to tadpoles.

Then climate change crossed her life path. Reflecting on this, Jann saw that her feelings of loss, grief, and anger about the deteriorating climate situation were understandable and appropriate, as the very notion of wildlife conservation—and thus her purpose—came into question. Now that Jann understood her emotional situation better, what next? How could she extend her eco-timeline? Should she try to recover meaning in her previous work, or chart a new path? The only certainty she felt was that she clearly needed to do something about her fatigue and sense of burnout.

Much like Larysa, Jann needed to rebalance the relationship between the big issues she wanted to address and her own well-being.

After years of wrestling with concerns about attendance at national parks, Jann decided to take a year-long sabbatical from public service to focus on restoring ten acres of property in a natural place that she loved. She wasn't giving up; she was redirecting energy to her personal sustainability and from large-scale conservation to caring for a piece of land on a scale that felt more intimate to her. She had ideas for the future brewing, but these could wait until she had the energy and motivation to take them on. For now, she needed to focus on her personal sustainability and renewal.

———

Much as is the case with environmental awareness, some people are born with an environmental identity, some develop it over the course of their lives, and others have it thrust upon them. Whichever is true for you, I want to empower you to knit your environmental identity to the other important parts of yourself, and consider how to use this knowledge to write the future chapters of your life.

Chapter 6

Values

I have spread my dreams under your feet;
Tread softly because you tread on my
dreams.

—WILLIAM BUTLER YEATS[1]

IN THE EARLY 1970S, RILEY DUNLAP, A GRADUATE student studying sociology at the University of Oregon, observed a controversy heating up in the nearby Willamette Valley. The local grass seed farmers' tradition of burning their fields at the end of the season had become a polarizing issue after the state began to regulate the practice as part of the new clean air laws. Citizens who were concerned about the negative health effects of air pollution were suddenly pitted against the farmers, who asserted their right to treat their land as they wished. Intrigued by this dilemma, Dunlap, who later went on to become one of the founders of the field of environmental sociology, began studying people's beliefs about the interconnectedness of human health and the natural environment—what he labeled the "New Ecological Paradigm."[2] At the time, the dominant and mainly unquestioned view in the industrialized world was one of "human exceptionalism": the belief that advances in science, technology, and culture exempted modern humans from the ecological constraints that applied to other species and earlier human societies. In other words, modern *Homo sapiens* was separate from and above nature.

In the fifty years since, the debate between proponents of an ecological worldview and human exceptionalism has only gotten more heated (pardon the pun). This clash of worldviews—the greater good versus self-interest—underlies much of our current environmental politics.

The reason this answer is so elusive is that it can't simply be found in facts, data, or theoretical frameworks; rather, it is in our *belief systems and values*. In this chapter, I'll introduce you to a number of tools and concepts that will help you understand different ways people balance their values about ecology and human welfare. You'll see your own values reflected, and it will give you insights into how others think, including people closest to you.

You first developed your values regarding nature as a child. They evolved through your education and life experiences along with your sense of environmental identity. Your values influence your moral principles of right and wrong and your *environmental norms*, the standards by which you judge your own and others' conduct in terms of nature and the planet.[3] But, as with any values, your eco-values remain unclear in the absence of language with which to express them. Once you can name the values you hold—for your own well-being, for others, and for Earth and other species—then you can celebrate them, embody them, and act on them.

"TREAD SOFTLY BECAUSE YOU TREAD ON MY DREAMS"

It's no exaggeration to say that examining environmental values is one of the most difficult and sensitive tasks we will explore in this book. Why? Your environmental values flow from your gut; they are a kind of core programming you have about what is right and wrong. You may not fully understand them, but you are likely to be sensitive about them. Your values are personal, even intimate, and they guide how you feel about controversial issues like endangered species, damaged landscapes, the plight of refugees, holding polluters accountable, what kinds of power sources should get subsidized by the government, or whether protesters should be punished for throwing soup at famous paintings.[4]

When I first started giving public talks to groups about sustainability and climate change, I would recite a poem by the Irish writer and statesman W. B. Yeats called "Aedh Wishes for the Cloths of Heaven." The speaker explains:

> ...I, being poor, have only my dreams;
> I have spread my dreams under your feet;
> Tread softly because you tread on my dreams.

Then I would remind the group to tread softly in our discussion, because in opening up about their deepest values regarding environmental and world issues, people were laying their dreams under our feet. That's how I feel about this conversation with you, reader, and how I hope you will approach similar conversations with others.

THREE BASIC ECO-VALUES

In the previous chapter, I asked you to consider how separate from or "at one with" the natural world you felt. How you answered that question depends in large part on the values you attach to nature. There are three basic attitudes that underlie a concern for nature or environmental issues: an ego-based view, which puts your own needs first; an altruistic view, which puts the needs of other people first; and a "biospheric" view, which extends your circle of moral concern to include other species and places.[5] These could be expressed as:

- *Egoistic nature values.* "I value nature because a healthy planet is good for me and my family."
- *Altruistic nature values.* "I protect the environment on behalf of the rights and well-being of others and for the greater good."
- *Biospheric values.* "Humans are one part of the circle of life. I value all creation—the birds and bees, the ocean and its marine life, the forest and trees."

While you likely identify with all these values to a certain extent, people often feel more strongly about one than the others. Larysa prioritized her own well-being and that of her children and family (egocentric values). For Marcus, the most pressing environmental issues were ones that affected other people (altruistic values)—specifically, those who lacked access to basic needs that others may take for granted, such as clean air and water, freedom to grow their own food, and safety when doing outdoor recreation. (Or, as he put it, "Hey, I love animals, but many dogs in this city have better housing, food, and health care than some people.") Reid, who had an awareness (even a sacred awareness) that he was part of a larger ecological system, chose biocentric values, citing his love of the ocean, surfing, and the stoke of catching a wave and being with the rhythms of nature.[6]

Sally and Martin

Salvatore, known as "Sally," was a client whose spouse, Martin, gave him a hard time at home over Sally's "obsession" with recycling. Sally was deeply concerned about the negative impact of plastics on the environment, from litter endangering marine animals to plastic nanoparticles leaching into the soil from water bottles and food packaging. I met Sally through the local Master Recycler program, a volunteer corps of community members focused on *sustainable consumption*: repair, reuse, composting, recycling, saving food, toxin reduction, climate protection, and environmental justice.[7]

Sally knew that while the rate of recycling was growing in the state of Oregon, where he lived, it was not keeping pace with the amount of new waste being generated. He believed plastic contamination was worse than climate change because it was invisible and insidious, happening inside our bodies, and this became the guiding principle of every purchase he made. Since it was impossible to avoid all plastic packaging, Sally strived to minimize buying new plastic by reusing plastic bags and containers, purchasing bulk items, and avoiding single-use products and nonrecyclable items. He set aside used paper and plastic containers for the local pickup, and diligently sorted lids, bottle caps, bread wrapper clips, and corks for a paid service that collected these other recyclables.[8]

However, Martin, an electrical engineer, was rarely mindful of the packaging in the items he purchased or how he disposed of containers. Martin was often annoyed by Sally's fastidiousness; Sally felt frustrated and alone in his daily battle against plastic, and couldn't understand Martin's apparent indifference. He was certain they shared similar environmental values. For one thing, they both enjoyed nature, being outdoors, gardening, and hiking. Martin had experimented with small solar panels and batteries to power his garage workshop and was researching new electric cars. Recently he had even started following the "citizen solar movement," taking on investor-owned coal and natural gas utilities that often threw up roadblocks against solar energy. So why did they have such different views on the importance of recycling? To help Sally and Martin navigate this, we had to go deeper on values.

A RANGE OF VALUES ABOUT ANIMALS AND NATURE

Some of the foundational research on environmental values was conducted in the 1970s by the late Stephen Kellert, then a Yale School of Forestry and Environmental Studies professor, at the request of the US Fish and Wildlife Service. Traditionally Fish and Wildlife had focused on the needs of hunters and fishermen, but in the 1970s its mission expanded to include the interests of hikers, backpackers, birders, and other "non-consumptive outdoor users." To effectively balance these priorities with other new responsibilities— such as enforcing the Endangered Species and Marine Mammal Protection Acts—they needed to better understand what this new breed of outdoorsman valued in the realm of preservation and wildlife.

Kellert's studies in the United States, Japan, Germany, and Botswana identified a number of values people held about animals, and by extension about nature, including strong relationship values with specific animals, such as one's pets; moral values such as opposition to animal cruelty or wasting resources; and utilitarian values that prioritize the practical benefits of exploiting animals and natural resources for human use.

Below is a list of value categories adapted from Kellert's studies. If you're involved in farming or hunting, you understand the utilitarian value of nature. If you train or breed animals, work in environmental engineering, or practice arts like bonsai, your values may include control and mastery of nature. I use a broader version of Kellert's original nature values, recognizing the importance of spiritual values and adding kinship values to better recognize Indigenous and First Nations views. Recently I included the value of survival to highlight life-and-death concerns brought by climate breakdown and other global environmental threats.

Reflect on which values stand out and feel like priorities to you. As with the three basic nature value orientations (egocentric, altruistic, biospheric), the categories are not mutually exclusive, and you can hold many of them at once. What gets interesting is the different *patterns* of values people have, the trade-offs we sometimes need to make to uphold certain standards, and how fractious it becomes when people's values don't align. (Note: The list is ordered alphabetically and does not imply a hierarchy.)[9]

- *Beauty / aesthetics.* Primary interest in the attractiveness and appeal of animals and natural settings, in particular charismatic species and iconic places.
- *Control and mastery.* Ability to subdue, tame, manage, or suppress nature and other species, ranging from training animals to redirecting the flow of rivers and even remaking entire continents.
- *Ecology and systems.* Valuing interdependence in relationships between humans, other species, and natural habitats; seeing the whole as greater than the sum of its parts.
- *Embodied experience.* Primary interest in direct contact with and exploration of nature and other-than-human species; active engagement with places, climates, and animals.
- *Fear and aversion.* Primarily negative views of animals or nature as threatening, unsafe, or unfamiliar; perception of nature or animals as unclean.
- *Kinship.* Focus on other natural beings and places as valued relations within the web of life, or as representative of one's clan or family.

- *Knowledge and science.* Valuing systematic observation and understanding of nature and other-than-human species, including underlying structures and processes.
- *Meaning and symbolism.* Valuing the natural world and other-than-human species for the ideals and meaning they represent, including in the arts or bringing good fortune.
- *Moral and ethical.* Reverence for right and wrong, order, and meaning in relation to nature and other-than-human species.
- *Relational.* Strong emotional attachment, caring, and love for nature and other species, including bonding and companionship.
- *Spiritual and transpersonal.* Value of nature and other-than-human species in relation to religious and spiritual beliefs and practices, or to the pursuit of wider levels of consciousness.
- *Survival and existential.* Fundamental concern about the ability to stay alive and meet one's basic needs, and trust in a future for oneself and one's family or community.
- *Utilitarian.* Focus on wise use and exploitation of animals and nature for practical human and societal benefits; seeing nature as subservient to human needs.

YOURS, MINE, AND OUR ENVIRONMENTAL IDENTITY: SALLY AND MARTIN

In our couple and family relationships, we have two or more environmental identities vying for space, acceptance, and growth. It's natural that friction will arise. Sally strongly identified with moral values about nature; he felt that polluting was wrong and acting to combat pollution was right. He appreciated the aesthetic beauty of the natural world, valued the experience of being outdoors, cared about wildlife, and valued his pets as fellow beings. He also valued knowledge and data-based climate science and even acknowledged the value of utilizing nature as a resource, like heating his home with wood and eating food grown from the earth.

When I had Sally rank these values in order of priority, he realized that some environmental issues affected him emotionally more than others. He also began to see how his values might differ from Martin's. Then I met with both Sally and Martin to explore where their differences lay, and what that meant.

After validating their different perspectives and explaining that the goal was to uncover new insights about themselves and their relationship, I had Sally and Martin work with a feelings list to explore the question "How do you feel about plastic waste permeating the environment, and all the health and ethical implications of this?"

Sally said he felt a rush of feelings: "anguished and hopeless," "ashamed" (about how his own actions contributed to the problem), and "outraged" (that plastic pollution was allowed to proliferate unabated). Martin was more measured. He acknowledged he sometimes felt angry and outraged too, but more frustrated and discouraged, and frankly somewhat resigned to the situation. When we dug in even further, it became clear that whereas Sally elevated moral values and the need to combat pollution by keeping plastic out of landfills, Martin put more weight on knowledge values. After all, he was an engineer, a technician. He derided the recycling system as a waste of time, phony "bread and circuses" to keep the masses happy. And he saw Sally's moral and caring values as sentimental—and ultimately futile, since Martin believed that most recycling ended up in landfills anyway. To Sally's surprise, Martin was able to describe in great detail why plastics were so ubiquitous, painting a picture of plastic as an industry, its connections to fossil fuel, and its marketing, convenience, and simple manufacturing process. Sally realized that Martin did care about this issue—he just felt powerless to do anything about it.

After opening up to each other, Martin and Sally were able to find a mutually agreeable solution: They would keep future purchases of products with plastic packaging (recyclable or not) to a minimum. Sally would continue his master recycling work and advocate at the state level for better laws about plastic pollution, while Martin committed to becoming more involved in the solar power movement in their city.

Questions about climate and the environment seem so big we don't think we can ask them. But you can ask people these things, and you need to. Romantic partners often have different climate beliefs and behaviors, and they are often inaccurate in their perceptions of their partner's climate beliefs and behavior.[10] This happens, in part, because many people assume that their partner believes and acts as they themselves do. Not surprisingly, partners' understanding of each other's beliefs and behaviors is significantly more accurate among couples that discuss climate change.

YOUR VALUES AND YOUR PET PEEVES

Sometimes it's easier to identify what you don't like than what you do like. If this is the case for you, try tapping into your environmental values by considering your pet peeves: issues that rile you up. Remember, we hurt where we care. Anything that makes you agitated and angry is a signal your values are affronted. Delphina, the veterinary student, and her partner, Gabby, didn't know whether to laugh or cry when they learned that hermit crab species around the world were adopting discarded bottle caps and other pieces of garbage to use as homes, instead of shells. Their ecological and moral values were affronted, as well as their value of caring for the crabs themselves.

Roman, the glacier scientist, was appalled by the selling of glacier ice that had been harvested from Greenland and sold for use in cocktail bars in the United Arab Emirates. He ranted to his fiancée about the irony of oil executives enjoying a glacier-cooled happy hour drink in an Arabian desert petro-state. For him, shipping glacier ice to the desert was a sign of control and mastery of nature that had gotten absurdly out of control. He couldn't just accept this as a legitimate use of glacier ice.

Jann, the Canadian wildlife biologist, was disgusted by the idea of "hunting derbies," with names like "Coyote Classic" or "Predator Palooza," in which hunters were encouraged to shoot as many coyotes or other species considered "varmints" as they could.[11] Having grown up hunting with her father and brothers (embodied experience values), she saw a place for hunting within sound conservation policy (scientific and ecological values). But these competitions were ugly to her and led to senseless overkill. In other words, they violated her aesthetic, moral, and utilitarian values.

For Agnes, one hot-button moral and ethical issue was the supersized carbon footprints of ultrarich individuals, particularly the celebrities who flew around the world in private jets, citing their own privacy or safety. As Agnes exclaimed, "What about the safety of the millions (or billions!) of people dealing with killing heat and drowning seas caused by global burning?" She wondered how her socially conscious friends could be oblivious to the "climate crimes" of their entertainment idols.

HOW YOUR VALUES KEEP YOU AFLOAT

I find people often want to skip over the values discussion and focus more on getting emotional relief or taking action. This is understandable. But values are one of the most durable aspects of your emotional coping and action process. Your moods and feelings will fluctuate given the circumstances, and what you consider to be the right actions will change and evolve (more on this in Part Five). You will undoubtedly face bad days, setbacks, and failures. Without connection to your underlying values, you are left with nothing to fall back on.

Healthy coping with environmental distress begins when you can reassure yourself that it is normal and expected to feel loss, guilt, or anger when your values are threatened. Positive feelings like pride and inspiration are important to express too! It's okay to feel great about making decisions that support your environmental values and make progress toward your goals.

In fact, your environmental values are a needed counterweight to your emotions. Picture a sailboat, with a heavy keel under the water that keeps it upright and tracking forward through the waves. Your values are like your keel, heavy and stable, always there to be relied on and to keep you going in the right direction. Your day-to-day emotions are like your sails, light and billowy and affected by all sorts of winds. On a good day, events may be in your favor and you will be sailing along. But during a storm, that keel is more important than ever to keep the boat upright. Other times you might lack motivation and have no wind in your emotional sails. During those times, you can remember your values, reset yourself, and start again.

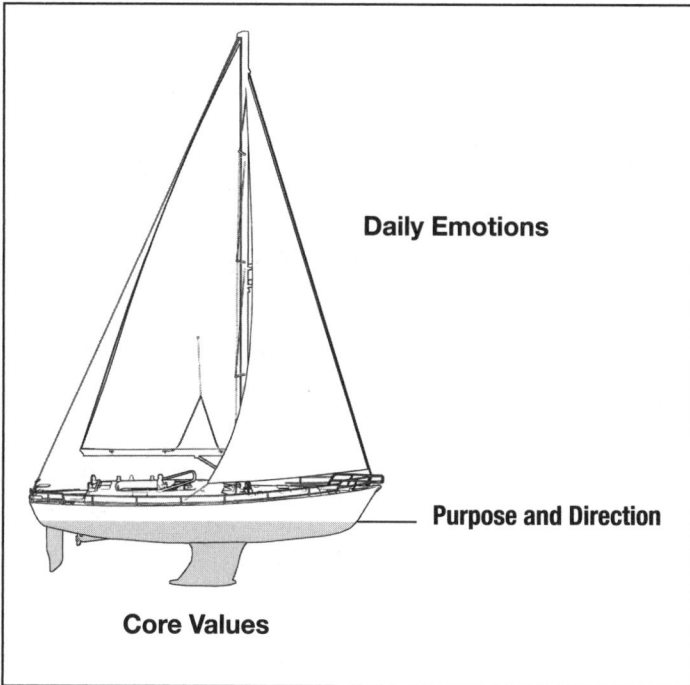

Daily Emotions

Purpose and Direction

Core Values

The keel image helped Delphina to orient herself no matter how troubling her findings were. She took solace in the fact that her fellow veterinary students shared the same values, so she didn't feel so alone. And when she had setbacks, got tired, or began to feel the normal overwhelm of multiple stresses (like the upside-down pyramid we discussed earlier), she could remind herself of her values and get back to her personal sustainability foundation.

BE THE CHANGE YOU WANT TO SEE

Having a language and framework for your environmental values helps you find (and recover) meaning in your environmental identity and eco-story, and can guide you in how to "be the change you want to see in the world." Most importantly, your values are building blocks for hope, determination, and personal sustainability. As artist and statesman Václav Havel reminds us, hope "is not the conviction that something will turn out well, but the certainty that something makes sense, regardless of how it turns out."[12]

Chapter 7

Nature

If you don't know where you are, you probably don't know who you are.

—Ralph Ellison[1]

CHRISTINA, A LIFELONG POLITICAL ACTIVIST, WAS AN EAST Coast chapter leader for the Citizens' Climate Lobby, a grassroots, nonpartisan group that urges US citizens to lobby their elected officials to take action on climate change solutions.[2] Her dark curly hair was now mostly gray, and she wore small cat-eye-frame glasses. Christina proudly described herself as a political animal, someone who thrived on being involved in government, elections, and policy.[3] She had a well-practiced public speaking style that helped her connect with people, often making upward-sweeping motions when tying ideas together, and then bringing her right hand down in a chopping motion onto her left palm when driving home her points. She had a knack for drawing in her political opponents, with a habit of gently clasping her hands and leaning closer, listening attentively, when someone else spoke.

Christina's climate advocacy had initially been inspired by a conversation with climate scientist Katharine Hayhoe. Soon after, she attended a training with Al Gore's Climate Reality Project and realized that the climate crisis was the most compelling political issue she could devote her life to.[4]

But despite her strong environmental voice and outward poise, Christina had a secret: She felt very disconnected and separate from nature.

When Christina came to me as a client, she confessed that she always felt like the young woman depicted in the Andrew Wyeth painting *Christina's World*, looking at the landscape around her but from a distance, always somehow apart from it. This was true practically—Christina spent very little time outside—but the distance was also psychological and even spiritual. And, she admitted, she felt self-conscious about her sense of remove from the outdoors, especially in the company of her environmental activist peers, many of whom had an insatiable appetite for heading into the woods. To address this *nature deficit*, I told her that it might be time to reexamine her assumptions about nature—and her place in it.

The concept of "nature" can be confusing in our technology-heavy modern society. It's easy to get lost in philosophical distinctions and semantics. With Christina, I used the metaphor of a pond or a small freshwater lake, not unlike Walden Pond in Concord, Massachusetts, made famous by the nineteenth-century nature advocate Henry David Thoreau.[5] If you visit Walden on a summer's day you can wade near the shore barefoot as small minnows dart around your feet, or join the people swimming in the water or gliding on paddle boards or canoes. However, not far from the shore, if you reach your toes deeper or dive under the water, you'll feel the colder layers underneath. Deeper still, at the dark bottom, are layers of sediment that can be read like tree rings, revealing layers of geological history.[6]

The pond metaphor, like the metaphor of the climate elephant, helped Christina see that there were different ways to approach her interactions with nature, and it was up to her to decide how deep she wanted to go. An insight that expanded her thinking came from mental health therapists who integrate healing work with nature and the outdoors (called ecotherapy). She could stick to the lake's translucent shallows, adopting a human-centered view of nature as an important health resource and protecting it as a pragmatic act of stewardship. Or she could wade into the lake's depths and mysteries, approaching places and species as sentient beings with their own rights and standing as more than just a commodity for humans to use.

In this chapter we will keep to the shallows and discuss how connecting with nature can support your environmental identity while also restoring your physical and mental health. In later sections, we'll wade in deeper to explore how you can find your own standing place, make a sincere land acknowledgment, and work through grief and avoidance of damaged nature.

VIEW THROUGH A WINDOW:
THE BASICS OF RESTORATIVE NATURE

In 1984, psychologist Roger Ulrich published a groundbreaking study, which found that patients recovered from surgery more quickly in a hospital room with a window view of green trees compared to patients with a similar room facing a brick wall.[7] In 1998, Frances Kuo and her team from the University of Illinois compared people living in identically designed housing projects in inner-city Chicago and found that in settings with green space there was less crime and inhabitants had more interactions with neighbors and a more positive outlook on the future.[8] In 2008, Peter Kahn at the University of Washington demonstrated that an actual window view has more benefits on heart rate recovery than a digital streaming view of the same image.[9] That same year, Stephen Kellert, whose model of environmental values we learned about in the previous chapter, published *Biophilic Design*, defining the now-popular architectural style that brings light, air, and a sense of proximity to nature into indoor spaces.[10] All of these studies and others demonstrate the benefits of nature for health and well-being.

When asked to describe a restorative natural setting, whether it's a garden or a glacier, people will invariably describe places that have some of the same characteristics: greenery, water, views, qualities of light and shadow, presence of other species. If you ask them what it *feels* like to inhabit that space, they will describe a sense of safety or comfort, an engaging bodily and sensory experience, and some balance between novelty, newness, simplicity, and familiarity. Such places are a source of beauty and fascination that effortlessly draw their attention, and they have a level of complexity that allows for deeper exploration and appreciation.

EXERCISE: Appreciating the Restorative Benefits of Nature

Take a breath. Be where you are right now, sitting, standing, reading, listening. Now, take a moment to visualize a place that you personally consider to be a safe, healthy natural setting. Allow yourself to bring it to mind. Maybe you have only ever seen the place in videos or photos, or maybe you have been there many times. Where is it, and what does the place look like and feel like? What do you hear? What do you smell? If you reach down to the ground, what do you touch? Soil, grasses? Maybe you are on the water. Or high up on a mountainside. Or in the desert. What do you see in the distance? What do you see up close? How does this place make your body and mind feel?

The feelings of calm and serenity associated with these natural settings are not just in your imagination. There are many proven benefits of spending time in restorative settings. These benefits fall into three basic categories:

- *Body.* Physiological benefits fostered by stress reduction, lowered heart rate and blood pressure, decreased cortisol and other stress hormones, and improved immune function.
- *Mind.* Cognitive benefits fostered by the surroundings and time for reflection, including a restored ability to focus your attention, gain perspective, and have creative insights.
- *Emotions.* Positive feelings fostered by beauty, self-reflection, enjoyment, relating, exploration, and play, leading to improved mood and increased self-efficacy.

You access these benefits of restorative places in various ways and combinations:

- *Passively*, just by having proximity to natural objects, whether indoors (with plants, an aquarium, or a window view of greenery), or outdoors in a safe, restorative setting.
- *Actively*, through outdoor exercise, nature appreciation like birding or photography; through yard, garden, or farm work; or adventure and sport.
- *Interpersonally and interactively*, through social activities with other people in nature, and interacting directly with plants and animals.

There is a "dose-response" effect for restorative nature, meaning that the more time you spend in your restorative natural setting, the more benefits you will enjoy. If you take a short break from work or study to get outdoors and get some fresh air, it will allow you to destress a bit and clear your head. But a longer break—like an afternoon or full day spent in your restorative place—can give you a physical and mental reset that will help you feel more refreshed. A few days or longer in restorative natural settings can really give you a sense of being away and allow for deeper reflection on your life. And extended retreats and journeys in nature can be life-changing. Typically, a good baseline to receive the minimal health benefits from outdoor time and activities is at least three hours per week.[11] This is similar to the common medical advice to get a minimum of three hours of vigorous exercise each week—and, yes, you can combine outdoors time and vigorous exercise to gain the benefits of both.

OVERCOMING FEARS OF NATURE

As a lifelong Catholic inspired by the Earth-centered spirituality of St. Francis, Christina had always dreamed of trekking the Via di Francesco, or the St. Francis Way—a historic Christian pilgrimage route near Assisi in northern Italy. The only problem was that she had no experience hiking, and now that she was getting older, she despaired that she wasn't mentally or physically fit enough to weather the 180-kilometer journey.

Christina's close friends weren't hikers, and she wasn't comfortable with the idea of going out on her own; she joked that if she were to set out alone in the wilderness, she would likely get lost, freeze, or "get attacked by wolves." Comments like Christina's are not uncommon when people speak candidly about their fears of nature, and are important to validate and to put in perspective. Your risk of getting killed by a predator—whether mountain lion, bear, or shark—is statistically lower than the odds of getting struck by lightning, and you are at greater risk of being killed on the highway driving to the trailhead than you are on the trail itself.

Despite her fears, it was clear that while Christina had a strong aesthetic appreciation of nature and highly valued the religious symbology of the Via di Francesco, what she lacked was the confidence to engage in outdoor activities, to keep herself safe, and to be comfortable in her own body. (These are all healthy developmental milestones for environmental education that I discuss in Chapter 8.) Christina had a fixed mindset that she was "not good" at outdoor activities and, by extension, "not good" at connecting with nature at all. Fortunately, connecting with the natural world is a skill that anyone can learn. I began by helping Christina see that "nature" wasn't something abstract and far away, but real and *nearby*.

FROM NEARBY NATURE TO WILDERNESS

To help her overcome her nature deficit, I showed Christina an image of a spectrum, where the natural world was accessible and present in various ways, including:

- "Virtual nature" like paintings, sculptures, and photography
- Indoor or "domestic nature" in her townhouse: her plants, her cat, her small yard and garden
- "Nearby nature" like the parks, creeks, and walking paths in her neighborhood and greater metropolitan area

- "Working nature" such as rural areas, farms, and woodlots in her region
- "Nature refuge" settings like resorts and camps
- "Wild nature" in the form of remote and hard-to-access places—mountain peaks, deep river canyons—or places legally designated and protected as wilderness areas

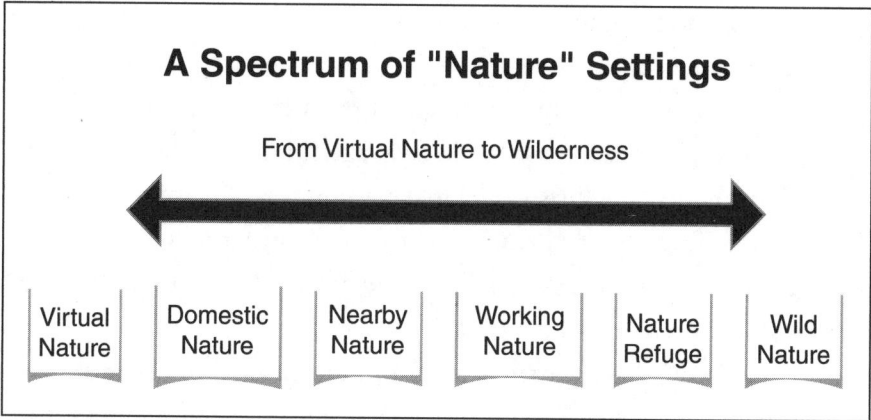

A Spectrum of "Nature" Settings

From Virtual Nature to Wilderness

| Virtual Nature | Domestic Nature | Nearby Nature | Working Nature | Nature Refuge | Wild Nature |

To just reduce all of nature to a few human-centric settings is overly simplistic. But these categories create *real places* to start. The spectrum helped Christina realize that, even as she dreamed of far-off adventure travel, she was discounting her access and opportunity to appreciate the presence of nature right in her home and neighborhood. Christina wanted to be more comfortable at the wild end of the spectrum; she now saw the incremental steps she needed to take in order to get there. The confidence she needed would be found not in the expensive gear she planned to buy for her future trek but in building the mental and physical capacity to experience and be present on the trek itself.

We focused on practical takeaways. First, Christina shifted some of her attention from her voluminous intake of two daily papers and online news scanning to learning the "news outside her door." I recommended a ten-minute sensing exercise each morning where Christina would simply take in the sounds and sights of her neighborhood, the fragrance in the air, the local bird life, and the background hum of human activity.

Next, she consulted the old analog field-guide books that had been gathering dust on her bookshelf, and augmented her knowledge of the names of animals and plants using some of the latest apps. I also directed her to stories of other women who confronted their fears of traveling alone outdoors such as Erica Berry's memoir *Wolfish*.[12]

Next, Christina found a local women's hiking group to join, an important step. I reminded her about the healthy synergy of doing outdoor activities in community with others, and about research showing that women specifically share outdoor and wilderness adventures together in a more bonded and noncompetitive way.[13] I told her the story of Lois Jones and her Ohio State colleagues, who became the first all-women team of scientists to reach the South Pole in 1969. They all put their feet down at the same time, to share the experience of being first.

Eventually Christina began a series of modest local adventures, trekking through different parks and trails, rain or shine, to build her fitness and learn outdoor skills. I gave her tips on how to layer her clothes for maximal comfort, how to track the weather, and how to pack what hikers call the "ten essentials," such as a map, water, a light, extra food, clothing, first aid items, and sun protection.[14] Christina was combining outdoor time with exercise and support from relationships with her group. These became bricks in Christina's personal sustainability pyramid.

Quickly, Christina felt a sense of growing familiarity and comfort with her local urban and suburban ecosystem. She didn't realize it at first, but in addition to her physical fitness, she was also building a growth mindset.

BEYOND NATURE DEFICIT

A sense of disconnection from the natural world like Christina's is not uncommon. Many people, in fact most people, in nations like the United States spend little to virtually no time outdoors; the Environmental Protection Agency estimates that Americans spend an average of 93 percent of their time indoors or in their cars. There is nothing wrong with being indoors, but there is something wrong with this imbalance.

Spending time indoors is a quite recent phenomenon in humans' evolution as a species, as technology has quickly taken precedence over fresh air, open spaces, other species, and clear skies. This takes a toll on your body and mind. Reversing this imbalance was the deeper process I worked on with Christina. She reminded herself that when she felt stressed or overwhelmed she could turn to her menu of nature settings to restore herself—fresh air and natural light, indoor plants, views of trees and greenery, the presence of pets and birds, gardens, outdoor activities and exercise, and mindfully taking on challenges and expanding her comfort zone.

Later, this process aided Christina's political advocacy work, by helping her realize that nature deficit is not just a personal problem; it's also a public health problem and an environmental protection problem. It's hard to get people to value the natural environment when it feels far away and abstract. As naturalist David Attenborough has said: "No one will protect what they don't care about; and no one will care about what they have never experienced."

A WORLDWIDE DIVERSITY OF NATURE STYLES

Your preferences and comfort level when it comes to spending time in nature can be very personal and culturally specific. For some, connecting with nature is individualistic, and solo experiences are ideal. For others, it brings to mind communal activities like family gatherings, meals, and picnics outdoors. People with experience-based values about nature may seek active recreation like outdoor sports or camping outings, while others are drawn to meditative and sensory experiences, as in the quiet practice of *shinrin-yoku*, or forest bathing, popularized in Japan and Korea.[15]

A classic study by Kathryn Williams and David Harvey highlights the value of what's known as "restorative-familiar" experiences: pleasurable and health-affirming activities we do in outdoor places we know well or return to again and again.[16] As a parent, I've made sure my daughter had a good mix of these experiences from an early age, making sure she could return to some familiar places to see how she was changing and growing with them. The Swedes have a wonderful term for a restorative-familiar setting: *smultronställe*, which literally means "strawberry patch" but is used to describe a personal or sentimental place you retreat to outdoors, whether your garden, a favorite walk, or someplace you love to visit.

If you are seeking what Williams and Harvey described as *deep flow* experiences, or being highly absorbed and in tune with an outdoor setting, you can embody *friluftsliv* (literally "free air living"): a Norwegian philosophy of life that values being outdoors and in the elements. Rachel, the city dweller worried about her daughter's choosing to attend college in a wildfire-prone state, had an adventurous side. When she was younger, she was a backcountry skier in Crested Butte, Colorado. This was a hidden source of confidence for her. Living in New York City, she found her nature fix in Central Park, tracking bird migrations and the acrobatics of the resident hawks.

Larysa, the nursing mother with a disaster-prepper tendency, dismissed the idea of being connected to nature because she didn't see herself as an "outdoor person." But after considering the spectrum, she realized that she had carefully designed her living space with organic energy in mind: preferring a minimal design with earth-tone colors, natural light and wood textures, and air plants arranged neatly on shelves. The spectrum helped her see that curating her home was a real way to embody her nature values.

Sally and Martin, the couple wrestling with their different environmental values, appreciated travel but were mainly homebodies. Entertaining friends and relatives in their shady backyard was a practice that grounded them, and the yard itself provided a domestic space in which to appreciate nature.

When Roman was first introduced to the spectrum, he felt only frustrated. The noise and bustle of his suburban setting was a sharp disconnect from the wild polar vistas he had become accustomed to. The reality was that Roman was still integrating the mind-altering awe he had experienced in the vast spaces and silences of Antarctica (what Williams and Harvey would call a *diminutive experience*), along with a startling new awareness of environmental crises (the "world of wounds" that Aldo Leopold spoke of). Looking again at the illustration of the nature spectrum reminded him he had options. While Roman couldn't replicate the frozen expanses he'd found in Antarctica, he could explore nearby outdoor places to find space and quiet to reset his nervous system as he considered his next career options.

STRUCTURAL BARRIERS TO
HEALTHY NATURE CONNECTIONS

As with other environmental problems, barriers to connecting with nature are both structural and behavioral, rooted both in society and in your personal beliefs and actions. Behavioral barriers can include a fixed mindset and lack of experience and practice outdoors, as was the case for Christina. The structural barriers can be physical, like the brick walls in Roger Ulrich's hospital experiment, or neighborhoods without access to parks, and they can be societal, as when racial exclusion denies you access to restorative natural spaces.

Marcus, the Portland city planner, joked about the dilemmas that recreational outdoor activities posed for dark-skinned men. When he drove out to a river canyon to go fly fishing, light-skinned anglers would either appear surprised and mildly uncomfortable meeting him on the riverside trail or treat him as some sort of exotic pathfinder, a fly fishing ambassador for people of color! Even in a city park near his home, there were legitimate dangers of being profiled as a threatening Black man, as happened to Christian Cooper in the infamous Central Park birdwatching incident in 2020.[17]

Lack of access to safe green spaces is properly seen as a *social justice* problem—one rendered more urgent and necessary in the face of threats like the dangerous heat of climate change.[18] Marcus was inspired by the urban landscape restoration work of Majora Carter and her famous "Greening the Ghetto" talk that raised consciousness about the degraded landscape of the South Bronx.[19] Marcus understood that you can't take safe healing in outdoor settings for granted; instead, you have to help create opportunities for it.

Braden

I first met Braden upon arriving at a Green Technology meeting in downtown Portland. I noticed him immediately because unlike the other attendees, who were mostly tech and urban planning people, he pulled up in a large pickup truck with an elk-hunting sticker on the cab. Braden was Oregon-born and had grown up in a family with long ties to the logging industry. He had come to the meeting to give a presentation about the benefits of new "mass timber" buildings being designed in the city (small skyscrapers with innovative composite wooden structures designed to be a strong, low-carbon alternative to concrete and steel).[20]

As a lifelong sportsman (whose family also supported conservation groups like Ducks Unlimited), Braden had environmental values that centered around hands-on, utilitarian activities in the outdoors, like timber harvesting, hunting, fishing, and training hunting dogs. Braden had a surfeit of confidence in his ability to survive in the wilderness. He had a lot of experience outdoors, wasn't intimidated by large animals, and believed that if an apocalypse came, he could live off the land. He also grew up as a Christian and saw himself as a steward of God's creation, though privately he believed that humans had overstepped their bounds in the natural world and, like any other species that grew too big for its habitat, were due for a "culling" from Mother Nature.

Braden considered himself to be more a doer than a talker, and this was his present challenge. He needed help articulating his values about nature and his vision for protecting forests in the twenty-first century.

ACCEPTING YOUR CONTRADICTIONS

Like many people committed to protecting humans, places, and species in the Anthropocene (myself included), Braden didn't fit neatly into one group. Braden loved the Earth and admired Indigenous peoples who lived in close harmony with the land. He had a degree in forestry and believed in science and innovation. He was an entrepreneur and a salesman. He understood enough about buildings and engineering to create timber-frame skyscrapers. But he was also old-fashioned. He idealized Oregon's heritage of logging and cowboy culture. As a child, he had spent summers at a ranch, and he still attended rodeos. In his office there was a framed print with the motto "There never was a horse that couldn't be rode, there never was a rider that couldn't be throwed." And though he had traveled the world, he felt most at home in small towns where he could see the horizon and stars.

Braden had a hard time navigating the politics of sustainability. As soon as people at the conference heard that he represented the timber industry, he felt judged. Yet when he was around some members of his family, he was teased for being an environmentalist and a tree hugger. He ended up feeling isolated, misunderstood, and resentful toward the environmentalists who assumed he was working against them, when in fact he was working to create wildlife bridges over mountain highways, an important conservation strategy for migrating species.[21] This drove him to work even harder on his business, which left him feeling fatigued and burned out. While his work and thoughts involved forests, he was spending more time looking at screens and sitting in meetings than out getting dirty, being around animals, and doing the things he loved.

To the outside observer, Braden's unique patchwork of interests and values didn't make obvious sense. But I helped him see that it made sense in the context of his life and his unique environmental identity. Over time, he found a new language with which to talk about his scientific, experience-based, and utilitarian values about nature, and like Christina, he became a better and more natural communicator. If he didn't fit the mold of what others thought a champion of nature should look like, that was okay. When it comes to your environmental identity and values, do you contradict yourself? To paraphrase Walt Whitman, yes, you contain multitudes.

CHRISTINA'S NEW WORLD

For Christina, the image of the brown-haired woman looking up at the landscape in the *Christina's World* painting signified a sense of helplessness and disconnection she sought to overcome by building her skills and being more confident hiking and spending time outdoors. When we learned the true story of the painting, the image took on a very different meaning. As it turned out, the artist Andrew Wyeth had modeled the painting's frail-looking woman after a real-life neighbor of his in Maine, Anna Christina Olson, who suffered from a degenerative muscular disorder.[22] Yet all accounts revealed Olson to have been a fiercely resilient person who refused to be confined indoors or in a wheelchair, and preferred to wander around the grounds surrounding her home, picking blueberries, and observing the natural world at close range on her hands and knees.

So Christina discovered a new, empowering interpretation of the painting: a person who was moving through the world despite limitations, and who was deeply in touch with the nature around her.

Another metaphor I love comes from the Greek myth of Antaeus, son of Gaia, the goddess who personified Earth. Antaeus was a formidable giant who challenged all who came his way to wrestle. And whenever he touched the ground, his strength was renewed, so he could not be defeated.[23] No matter where you live, you can "touch the ground" by partaking in the vast spectrum of what nature has to offer in whatever way you choose. One person walks daily in Central Park, another goes on a months-long pilgrimage, while another hunts and fishes. What about you? How do you aspire to deepen your sense of connection with nature?

Chapter 8

Family

We have such a brief opportunity to pass on to our children our love for this Earth, and to tell our stories. These are the moments when the world is made whole.

—RICHARD LOUV[1]

THERE'S AN ENVIRONMENTAL IDENTITY EXERCISE I LIKE TO do with groups of people of all different ages, ranging from folks in their seventies to younger teens. I ask them to line up in order of the age at which they first *really* grasped the gravity of climate change. People are often confused by the question at first. Then they start animatedly talking among themselves, comparing details of their stories, until eventually they arrange themselves into a line.

Next, I ask people to share why they placed themselves where they did. When the eldest of the group speak of their "waking-up moments," some harken back to the first Earth Day in 1970 or to reading books like Rachel Carson's *Silent Spring*. They link their awareness of climate change with their general consciousness about ecology, usually developed well into adulthood. As we start to move down the line, people point to seeing Al Gore's *An Inconvenient Truth* as young adults in the early 2000s. As we get closer to the present day, some talk about how they learned about climate change in college or high school, often naming disaster events that raised their consciousness. When we get to the youngest of the group, they say there isn't a time they weren't aware of climate change. It's the only world they know.

An exercise like this is a great way to illustrate how environmental awareness is keyed to different stages and events over the course of your life, and how it differs from one generation to the next. As an early Gen Xer, I got through college without ever hearing the words "climate change." It was only in the late 1990s, when I was in my early thirties and I sat in on environmental studies courses during my psychology training, that I got a sense of the physical science of "global warming" (as it was then called) and how it affected the whole of Earth's systems and biogeochemical cycles.[2] For young people, it can be a surprise to learn there was a "before time"—a time when you didn't have to know about things like climate change.

The oldest living people wouldn't have heard climate change discussed in popular culture until they were in their sixties. On the other end of the spectrum, members of Generation Alpha, the first fully twenty-first-century cohort, who are now reaching their teen years, have studied climate change as part of their school curriculum since a very early age. Some young people are well-known climate activists themselves.[3] And as I write, Generation Beta (those born after 2025) is joining us, the first generation who will live to see the twenty-second century. They will come of age in the future we evoke when we consider the next century's global temperature and sea level rise scenarios.

The Yale University and George Mason University research programs on climate change communication have found that an attitude of alarm, concern, or caution about global warming is reported by a majority in every generation in the United States and in most countries worldwide (we'll look at this more detail in Chapter 9).[4] And these fears can shape some of life's most critical decisions, from youthful wondering about what life goals to pursue, to whether to have children (and how to parent, teach, and mentor them), and to how we plan to spend the later years of our lives.

While these fears may know no generational bounds, those who have lived the bulk of their life without having to wrap their head around realities like climate change have a qualitatively different way of approaching the issue than those who have been born into it. This manifests in different ways. Gen Alpha was born into a sense of possibility: They are the most materially endowed, technologically interconnected, and likely the longest-living generation ever. But at the same time they have grown up in a world of ever more dire environmental warnings and insecurity, and will be the most threatened by tipping points in climate and biodiversity across their lifetime. For some elders, having lived such a large portion of their life in the "before time" can contribute to blind spots and difficulty taking on the new reality. But others feel a sense of responsibility to future generations and struggle to assuage the fears of their children and grandchildren. How can they transmit their hard-earned sense of resoluteness that comes from having lived through challenges that younger people have not yet developed?

So we need to learn from each other.

Take a moment and think about your generation and what age you were when climate change—or perhaps the whole cluster of wicked problems of which climate is one part—became real to you. When doing this exercise, some people admit they are just now becoming aware of the gravity of the climate crisis. That is a perfectly appropriate answer too.

ENVIRONMENTAL IDENTITY AND YOUR FAMILY

Therapists like to joke about the client who comes into the first meeting and says, "No family stuff, no childhood shit, I just need some strategies." But therapists know that it's impossible to fully disentangle your family and your childhood from your adult struggles, including eco-anxiety and grief. They also know that while you don't need to get stuck in your past, it helps to understand it.

For Rachel, the therapist who was secretly terrified about her daughter, April, going to college in Southern California, the reluctance to share these fears could be traced back to her complicated relationship with her own mother. Ingrid had been a fairly hands-off Colorado hippie parent who encouraged her children's independence. Rachel had once appreciated her "free-range" childhood, but upon becoming a parent herself she realized that her mother's laissez-faire parenting sometimes bordered on neglect, and so she vowed to have a different relationship with her own daughter. As April prepared to leave the nest, Rachel was having a hard time balancing her impulses as a parent. Should she protect April in ways that Rachel herself had not been protected? Or should she let go? Rachel needed to make peace with unresolved feelings about her mother so that she could be present with her own daughter and have an honest conversation about her fears for April's safety.

This is an example of how *family dynamics*, the unique relationship patterns and communication issues that exist in families, and our *psychodynamics*, the unconscious ways early experience affects us in the present, play a big part in the development of our environmental identity. In this chapter, we'll explore how family relationships form the foundation for your environmental values and beliefs and how you express them—whether by continuing your family's tradition or charting new paths. And we'll discuss how, if you are a parent (or considering becoming a parent) yourself, the family you create will be the primary training ground for your children and grandchildren to develop their own values, identities, and choices.

ENVIRONMENTAL IDENTITY AND WELL-BEING: RISKS ACROSS THE LIFE SPAN

The emergence of an environmental identity—and the potential to encounter issues like eco-anxiety—manifests differently at different ages (though this general development trend will have personal and cultural variations).

Babies experience a pre-language world of senses and movement; they seek pleasure, comfort, and safety from caregivers. As the child's first source of information about the world around them, caregivers can shape a child's earliest impressions by providing safe ways to explore, touch, and taste the natural world—and by being conscious of transmitting negative, fearful messages about nature, Earth, and the outdoors.

Young children have a natural openness and interbeing with nature and other species and a tendency to blend make-believe and the real in their play.

They also can have normal anxieties that might be influenced or heightened by an awareness or experience of disasters or losses, especially in the case of precocious learners who may discover material on their own before they have the coping abilities to handle troubling news. Naturally egocentric, young children can blame themselves for society's problems. They may lack the language to express their concerns and act them out instead.

Adolescents are gifted with powerful new reasoning abilities and an expanded perspective that help them see the world in new and innovative ways. They also have to integrate new awareness of eco and climate issues while developing their identities in a highly fluid culture (to which online social dynamics adds additional layers of complexity). For some, the gravity of eco and climate issues can produce additional anxiety on top of existing stressors from school, peers, sports, academic pressures, and family relationships. Idealistic young people may take on projects that set them up for hard lessons about the difficulties of making social change.

Young adults can take control of their own environmental identity and assume the freedom to manifest their dreams. They are also expected to leave the nest and create a life vision for themselves in a culture that actively questions its ecological future. Coming of age in the midst of environmental crises can contribute to an empowering sense of urgency and commitment, or to moments of alienation and hopelessness. Involvement in activist groups can bring its own pressures for identity and performance.

New or prospective parents must manage dreams, obligations, family traditions, and ethics about childbearing in the era of climate change. Parents' normal concerns about the well-being of young children, and the stress and fatigue from balancing personal, family, and work responsibilities, are compounded by existential concerns about climate disasters and a livable future on the planet.

Working adults can find that environmental and climate disruptions affect their economic, occupational, and career prospects. Environmental or conservation professionals and disaster responders must bear the additional toll of burnout, trauma, and physical danger that comes from being exposed to climate catastrophes.

Elders are challenged to maintain their physical and mental health and manage multigenerational family relationships in a time of environmental upheaval. Celebrating one's values and legacy can be a joy. Questioning legacy can be a burden. What world are we leaving for future generations? Longtime environmentalists have to grapple with the potential for despair as they face renewed setbacks.

Jesse

Jesse, a fourth-generation resident of western Oregon, sought me out for grief and family counseling after catastrophic wildfires destroyed large tracts of her family property and tree farm. This was the treasured landscape of her childhood: vast hillsides of fir and cedar, alder and hemlock; misty forests, creeks, and valleys that stretched as far as the eye could see. So Jesse was devastated when the homestead that had been in her family for five generations burned to the ground, leaving just the charred old river-stone fireplace standing starkly among the ashes.

Jesse, her siblings, and her cousins shared ownership of the property, and they had never been exactly in agreement about stewardship. After the fire, the disagreements came to a head. Some siblings wanted to let the forest restore itself naturally. Others wanted to use the fire as a rationale for "salvage logging"—skirting regulations to cut down the remaining charred trees to recover the financial value of the land, and cutting healthy sections of forest to prevent other blazes. While they spoke of the same goal—to restore the land—they approached it from different value sets. Reaching a compromise required difficult decisions that caused rifts between family members.

Jesse was opposed to the idea of trying to profit from the land in the short term, preferring to place it in a conservation trust so that it would be preserved for the future. To her, salvage logging was one of the tentacles of global capitalism with its endless commodifying and profit-seeking, especially unsavory in the aftermath of disasters (what's known as *disaster capitalism*). At the same time, she understood that members of her family were relying on the financial value of the land to ensure their families' well-being.

Beyond the short-term dilemmas, Jesse struggled with the question of how to honor the Native people who had made their homes on the land before her family arrived, while still expressing her own emotional connection to the land. (We'll look at the complexities of our place connections in Chapter 12.)

I recognized Jesse's environmental sensibility as that of a nemophilist (literally, "lover of tree groves")—someone with a profound fondness for forests, woods, or woodland scenery. Her attachment to the living forest ecosystems of her youth was integral to who she was as a person; she often felt like the trees she had grown up with were like an extension of her own body. Her basic environmental values orientation was biospheric; the circles representing her self and nature overlapped. In contrast to Braden, the timber harvester, who also felt intimately attached to the land but held utilitarian values, Jesse believed that harvesting the materials we need to live from forests was morally acceptable only after determining what was good for nature as a system.

THE ECO-FAMILY TREE

I envy people like Jesse and Braden with lifelong connections to a beautiful natural place—even if that place has now become vulnerable in an era of unpredictable climate disruptions. If I were to sketch out my family tree from an ecological perspective, it would look very different from Jesse's. She would have five generations on the same mountain, pursuing the same way of life that was connected to the land. Because I grew up in an old industrial rust-belt city, Buffalo, New York, my relationship with the land was very different.

My mother and father grew up in the 1940s and 1950s a few blocks from each other in a traditionally Irish and Polish working-class neighborhood in south Buffalo. The neighborhood was called "The Valley" because it was surrounded on all sides by steel mills, grain elevators, and rail lines. Five iron bridges offered the only ports of entry, and the pristine nature that had once existed in that area near the fresh waters of Lake Erie had been erased with the construction of the Erie Canal in the 1820s.

When I was new to the concept of environmental identity, I approached it in a black-and-white way: Either you had an environmental identity or you did not. Back then, I would have put my parents in the second category. But today I realize that this was a false or at least limited understanding. My parents and I had *different* environmental identities and *different* senses of place. Whereas I consider myself highly attentive to the natural world and my local ecosystem, my parents mainly associated nature with manual labor, such as shoveling snow, cutting grass, or raking leaves (utilitarian, mastery, and survival values).

And while we never spoke about attachment to place, my parents certainly had one. They both lived their entire lives within a few miles of where they were born: the same streets, the same climate and weather, the same churches and businesses. My attachment to place is shaped by my education and travels and now includes many places. One approach is not necessarily better than the other. The reason these ideas matter is that they give us a new language to talk about our relationship with nature, and ultimately the planet.

DRAWING YOUR ECO-FAMILY TREE

Imagine drawing a family tree diagram spanning multiple generations: you, your parents, grandparents, spouse or partner, siblings, children, grandchildren. Then, consider how each person's place of birth, the era they grew up in, and the life paths they pursued shaped their environmental experience and values.

This will require some reflection, and also some investigating. Think about aspects of the living world that shaped your family history, such as migrations, the influence of world events, and disasters that left their mark on the generations. Ask elders to share their stories, and ask younger people about their relationship with the natural world. While having these conversations, however, it's important to be mindful of sensitive areas. Some branches of your family tree can carry loss, conflicts, trauma. (These are signals and opportunities for more caring and healing conversations, on your own or with help.)

In the same way that you used the eco-timeline to track the development of your environmental identity, you can use an *eco-family tree* to visualize how different environmental identities developed over the course of your family history. Thinking along the lines of other generations' environmental identities and values will expand your understanding of where you have come from: the environmental legacy you will pass down to your own children and to future generations of your family.

GOOD-ENOUGH PARENTING IN AN ERA OF CLIMATE CRISIS

It was March 11, 2011, the late afternoon on the West Coast. I was sitting at my living room table scanning the news on my laptop before dinner. I was engrossed in live coverage of the Tohoku earthquake and tsunami in Japan, which sent ocean waters surging through the coastal cities. At that point, with the disaster still unfolding, we didn't realize that Tohoku would turn out to be the most powerful earthquake in Japanese history and the fourth-most-powerful tremor ever recorded. Nor did we foresee the reactor meltdowns at Fukushima that would transform this into a na-tech disaster (a technological disaster triggered by a natural hazard).

My daughter, who was about three and a half at the time, crawled onto my lap, and her eyes followed mine to the screen. She asked me what I was looking at. While I sat engrossed, I explained that huge waves from an earthquake deep under the sea were flooding cities in Japan, a place on the other side of the ocean from where we live. She watched for a moment, then looked up at me and asked: "When will the tsunami come to Portland?"

As a psychologist, I knew it was perfectly normal for her to ask that question. Young children see the world in relation to themselves, their family, and where they live. That moment of watching the unfolding Fukushima disaster in real time did not scar her for life (in fact, she doesn't even remember this day). But it was a wake-up call for me as a parent.

Watching traumatic events in real time—even if from a distance—can scar people emotionally and can even cause post-traumatic stress disorder,

an effect that was well studied after the 9/11 attacks.[5] My daughter's innocent inquiry reminded me that I needed to be more aware of what information she was taking in—and remember that, as a parent, in shielding her I
was absorbing the damage.

Developmental psychology—an approach that plots our social, emotional, and intellectual growth by the stages of our lives—teaches that perfection cannot be the goal of parenting. The goal is to do your best and
be "good enough." But what is "good-enough parenting" when it comes to
answering the questions that children ask about climate crises and other
environmental disasters?

SUPPORTING YOUNG PEOPLE
IN THE ERA OF CLIMATE CHANGE

Several years ago, I visited my daughter's fourth-grade class to talk about
climate change and the basics of climate science. I brought two props: a
basketball and some plastic wrap. Using a desk lamp as the sun and the basketball and some other round objects from the classroom as the planets,
we made a model of the solar system to demonstrate how the sun brings
light and energy to Earth. Then we wrapped the basketball in plastic wrap
to show how our thin layer of atmosphere traps the heat that creates a livable climate for life (it may sound simplistic, but modeling Earth's atmosphere with the basketball and plastic wrap is fairly true to scale). To
demonstrate how the atmosphere might feel to Earth, I had students cover
their arms with the plastic wrap and notice how it trapped their body's heat
and skin's moisture. We talked about how the atmosphere is really quite
thin and fragile—only sixty miles high on our 8,000-mile-diameter Earth,
equivalent in thickness to the skin of an apple!

Environmental educator David Sobel is known for the saying "No
tragedies before grade 4."[6] By that he means that we should avoid saddling
children with difficult and developmentally inappropriate environmental
knowledge. Instead, we should teach them in age-appropriate ways in service of building a foundation of healthy attitudes about themselves, nature,

and Earth. This will prepare them for when they encounter more complex environmental issues later on.

The physics and chemistry of climate change are not hard to grasp. To understand climate-related concepts, all kids need is hands-on information showing how it would be possible to change the gases in the thin atmosphere of Earth and what that might mean for life on the planet. Children learn best from visual metaphors—snow sliding down a sloped roof becomes a model for a glacier, a basketball and some plastic wrap becomes a lesson on the atmosphere and heat-trapping gases. Discussions about weather are more meaningful when children are outside actually looking at the sky (and stopping to find shapes in the clouds). Rather than learning bird species from a textbook, young children can dress up like birds and learn about how they live through acting and playing. And when we give them the chance to go outdoors and practice exploring one square yard of soil with some stakes, string, and a magnifying glass, counting all the species they find there (ants, earthworms, etc.), we invite children to become attentive to the process of life. In every hands-on lesson, children gain a small sense of confidence and mastery.

What I am describing is called developmentally appropriate environmental education, a play-based approach to teaching about nature and ecology that is tuned to the cognitive and emotional needs of children as they grow. An age-appropriate scheme for teaching would include:

- *Ages four to seven.* A focus on developing a sense of wonder, teaching healthy feelings, and nurturing a sense of connectedness with the natural world through stories and songs, play, and moving like animals.
- *Ages seven to eleven.* Foster self-efficacy through hands-on exploration and immersion in the natural world: exploring streams and pathways, building forts, taking care of animals, and gardening.
- *Ages eleven through fourteen, and beyond.* Impart knowledge and abstract concepts about how systems in society and the

natural world work, including questions of ethics. Keep it local with school programs and environmental questions and activities centered on the student's home place.

Developmentally appropriate environmental education can be a family affair. As Louise Chawla, one of the eminent researchers at the intersection of environmental psychology and child development has revealed, the childhood experiences that best teach understanding and responsibility for nature include a combination of many hours spent outdoors in a keenly remembered wild or semi-wild place, and also time spent with an adult who teaches respect for natural processes. She highlighted how "joint attention"—a child and parent (or other adult) paying attention to the same thing, intentionally—fosters the child's nature values, while also strengthening the adult-child relationship.[7] Think of this as a "three-way attachment," creating healthy bonds between a child, an adult, and nature simultaneously.

TEACHING NATURE VALUES TO YOUNG CHILDREN

I remember once gardening with my daughter. She was maybe two and a half. While digging, we discovered a plump earthworm. She picked it up, looked closely at it, and then promptly pulled it in two.

Some parents would respond to this incident with concern or even horror. But rending the earthworm is a normal response for a curious young child. Incidents like these are teachable moments: opportunities to help a child understand that other species can experience pain and suffering.

From a very young age, we all have *instinctive* responses to the natural world. We are evolutionarily wired to experience spontaneous feelings of love and affiliation with nature, known as *biophilia*, just as we are primed to react quickly to certain threats from nature, such as spiders, snakes, being at dangerous heights, or being in dark, enclosed places. Both sorts of reactions are evolutionary adaptations necessary for survival.

We also have instinctive *moral* responses. Babies and young children develop a rudimentary moral sense from the very start of life. They show empathic responses to people and animals and an impulse to soothe hurts. But actual *awareness* of moral choices comes later, as kids learn from what they observe in role models around them. My daughter was not born knowing about the morality of hurting creatures like worms; it is something she had to learn.

As parents, teachers, and mentors, our job is to take the basic evolutionary preparedness that children come into the world with—their instinctive and moral responses to Earth and other-than-human species—and to help nurture it through information and knowledge. Our role is to help an innate, partially formed environmental identity develop into a mature and healthy one.

SUPPORTING TEENS: MEANING-FOCUSED VERSUS PROBLEM-FOCUSED COPING

Research shows that today's teens feel "stripped of power," "stranded by the generation gap," and "daunted by the future."[8] Being up-front and honest when talking to teens about climate and the physical world helps demystify the facts. Teens are moving into adult developmental territory. They can develop a full, 360-degree range of feelings and a growth mindset, and they have untapped potential for greater capacity and learning. When we engage with them honestly, the explicit message is "You can understand this."

Young or old, everyone needs to know that their concerns are valid. Yes, my client Agnes, the teen activist, is right. The world she's been given is full of increasing threats that are not of her generation's creation. Her sense of despair over climate change needs to be validated and elevated, even when she's grappling with other issues as well. Too, she needs to learn that while there is value in actions, there is also value in understanding the meaning behind those actions.

According to Swedish psychologist Maria Ojala, both *problem-focused* and *meaning-focused* coping strategies are helpful in preserving psychological well-being. Problem-focused coping targets immediately actionable responses: doing a beach cleanup, creating a recycling program, planting a garden. It's most effective with specific challenges that have a clear and viable solution and require action to achieve the outcome.

Meaning-focused coping involves reflecting on the significance of your actions, including how they align with your values and long-term goals. It's more effective for larger and diffuse problems for which there isn't one obvious solution and which require long-term engagement and commitment. When a problem can't be solved, the meaning of our attempts gives us something to fall back on (as do our values).

I teach teens (and adults) to toggle between action- and meaning-focused strategies.

CROSSING PATHS

Climate change and environmental problems don't only impact our individual psyches; they also affect our *family dynamics*—how family members influence each other's thoughts, feelings, and behaviors. There's a process in therapy informally known as "crossing paths," when one family member enters a new life stage and this activates unresolved issues on the part of other family members.[9] This happens with environmental identity. For example, a young person who has been educated about climate issues and begins to express strong environmental beliefs and values might bring her parents and grandparents to a necessary point of self-reflection regarding their own environmental beliefs or values.

Young people stricken with grief about massive ecological losses can't imagine why others are able to move through the day seemingly unburdened. Agnes looked at her parents and thought, How can they not be freaked out by this? How can they not *act*? With some mentoring, Agnes was able to realize that not everyone in her life was going to prioritize the same problems, and that a parent focusing on keeping family and career going may not—cannot—focus as much mental energy on thinking about the melting of faraway polar ice caps. They may share her goal for a healthier planet, cleaner environment, and arresting global warming, but their bandwidth to engage with these issues may differ.

COUPLES, COMMUNICATION, AND THE CHILD QUESTION

We saw earlier with Sally and Martin how couples can become divided about mundane issues like recycling and how, when left unaddressed, this can result in hurt, anger, and misunderstandings. Roman and his fiancée, Sandra, found themselves mired in an even bigger conflict after Roman returned from his season in Antarctica.

To set the stage: Roman had difficulty with emotional expression because this hadn't been modeled in his childhood (a little-*i* issue). He was also struggling with eco-anxiety and ruminating about the world and the future (a capital-*I* issue). These issues came to a head when Roman disclosed to Sandra that he was considering getting a vasectomy so he would not father children; as he saw it, to bring a child into the world was to be complicit in the global population growth that was depleting Earth of its natural resources while contributing to carbon emissions.* He'd been keeping this to himself because he knew Sandra was considering the possibility of having children and he didn't want to hurt her feelings or jeopardize their relationship.

I told Roman that I understood the mix of feelings he identified (confusion, guilt, shame, fear of conflict and of losing Sandra, anger at the state of the world). The question of whether or not to have children can be highly fraught because it's wrapped up with our dreams, the expectations of our family, cultural and religious obligations, innate biological urges, our personal health, and of course the responsibility for the life of a child and adult of the future. And in Roman's case, ethics about nature and ecology featured prominently on this list.

Then I helped Roman to include Sandra in the discussion.

We started, as with all couples therapy, with basic communication styles and managing the fight-or-flight response. Roman preferred to focus on the positive and to use humor when uncomfortable emotions came up. I asked him if he shared his deeper feelings with Sandra and he said, "No, not typically." Sandra admitted that when she wasn't sure what Roman was

---

* Several of Roman's Antarctic research colleagues had done this, as a statement about global population and human impacts on the environment.

feeling, she would "pick a fight." I applied a fire metaphor I use in therapy for expressing emotions. In nature, wildfire is a natural and healthy part of the forest ecosystem, and sometimes foresters start fires deliberately, purposely burning underbrush to reduce fuel for future fires in what is known as a *controlled burn*. I suggested that maybe when Sandra was "picking a fight" she was hoping to start a controlled burn, in order to avoid a bigger, forest-razing explosion. I suggested that Roman and Sandra's communication patterns likely came from their families of origin. Sure enough, Roman joked, "My childhood was an arctic tundra, nothing to burn. Cold, Germanic, never saw a fire!" Sandra's Argentinian family was the opposite. Every emotion was exposed for every other family member to either douse or inflame.

Roman and Sandra were not alone in weighing eco and climate issues when considering whether to have children or to grow their family. Having seen this before, I thought they could benefit from a visualization tool I use to give people a wider perspective on their feelings and decisions. I had them imagine they had an *eco–time machine* that would allow them to travel to the past. No matter which era Sandra or Roman chose, they saw that throughout history, in times of existential threat such as war, economic depression, and fears of nuclear destruction, people have questioned the wisdom of bringing new lives into the world. There was no one right or wrong decision.

Taking a cue from climate researchers like Jade Sasser and Britt Wray, in addition to questions like "Should we have children?" I encouraged them to consider "What's required to have a child today?" and "How do we support children in this world?"[10] And what if they let go of the assumption that procreation was somehow the most important vote they could make about Earth's future, when in reality there have always been many roles for people to play in healthy societies?

Sandra shared Roman's view that climate breakdown was an existential threat. They needed a way of expressing how they were thinking and feeling that included their environmental values and identities. So I had them each create an eco-timeline of their life and then use the time machine to visualize ten, twenty, or more years into their future. Then they could compare their visions for their life path and imagine how they might grow their eco-family tree.

It was still not easy, but this groundwork helped them communicate with more compassion and openness, which ultimately helped them feel better about their decisions. Later, I also helped Sandra and Roman see that taking on the child question solely as one isolated couple was yet another form of the *climate hostage* situation, where we are expected to make decisions to contribute to the future of society without being supported by society, and even when the evidence we see makes us question a future at all.

FILLING IN GAPS IN YOUR DEVELOPMENT

Psychologists define healthy psychological development across one's life span as the integration of qualities and abilities first forged in childhood with the experience and earned wisdom of adulthood. Here are some examples of attributes that make up a mature environmental identity:

- Connection with your body and senses (including ways to remain physically fit)
- Sense of awe and wonder (including an ability to play and imagine)
- Competence in outdoor and nature-directed activities (camping or living outdoors, gardening, hunting and gathering, dealing with varied weather)
- Ability to grasp complex ecological subjects (including ethics and committing to action)
- Fortitude and determination (the ability to stick with projects over time, but also to let go of outmoded beliefs or strategies)
- Patience, compassion, and wisdom (and the ability to share these qualities with one's family and community)
- The ability to transcend or think beyond one's individual life (in terms of one's own mortality, others, and the more-than-human world)

Lest you think that psychological maturity requires leaving your childhood self behind, I want to emphasize an important aspect of healthy development that I have hinted at. It's never too late to have a second childhood. Childhood play and wonder have their adult manifestations and are possible at any age. In order for you to help children and young people balance their instinctual fears about nature with appreciation and awe of its beauty and splendor, you need to keep working this out yourself. This is where all of the thinking, feeling, de-stressing, risk assessing, connecting, and identity-making work you have been doing comes into play.

Though no one can embody these qualities all the time, it's helpful to keep this vision of healthy development in mind. We will return to many of these qualities in subsequent sections, as tools for healing and flourishing.

PART THREE

Healing

WHEN THE EMOTIONAL DISTRESS BROUGHT ON BY CLIMATE and environmental threats leads to serious or debilitating anxiety or depression, you need more specialized tools for self-healing. We'll begin by exposing how polarized politics and climate propaganda help drive your eco-distress, and we'll take a tour of positive and healthy approaches to political engagement. You'll learn about the personal and cultural faces of eco-anxiety, how worry and fear about planetary stresses affect you, and how to restore anxiety to its proper role in your spectrum of emotions. I'll share ways to tend to the eco-grief and depression you may feel, recover your energy, and reengage with life. Next, we'll look deeper into your connections with nature and learn how to restore your ability to enjoy outdoor time when the places around you themselves become damaged, as in times of deadly heat, poor air quality, and amid the scars of disasters. When you build capacity to cope with issues like anxiety and grief and learn to care for damaged nature, you are ready to find your own *standing place*, a piece of land you cultivate an attachment to and commit to protect.

You may think that your mental health coping skills are only useful for problems inside you, or within the small circle of your relationships, family, and work. However, the lessons of mental health therapy and self-help have much to offer for surviving climate anxiety. In fact, mental health coping becomes a supercharged *ecotherapy* when we add what you have been learning about your environmental identity and values, the connections to nature in your daily life, and your eco–family tree.[1]

Chapter 9

Hostage

One day, when it's safe, when there's no personal downside to calling a thing what it is, when it's too late to hold anyone accountable, everyone will have always been against this.

—Omar El Akkad[1]

In 2018, I coined the term *climate hostage* to capture the powerlessness and frustration I was hearing from my clients and people around me.[2] It was some version of "I feel trapped in a society where there is so much wealth and potential, and we all realize the urgency of the climate crisis, but somehow denial and inaction pin us where we are, on a planet hurtling toward disaster, with no possibility of escape." If you feel this way, it's understandable. When you look at the news, a state of worsening ecological breakdown can feel like an unchanging reality you must endure. But when you look closer, you find some things can and do change. Not *everything* you may want to do to solve eco and climate problems is possible, but some things are.

When it comes to climate anxiety, "the personal is political."[3]

We need to look at politics first if we are to understand eco and climate distress. By "politics," I don't just mean whom you voted for in the last election; I mean how you exercise your power and agency as a person, and claim your human rights and your place in society. This includes your beliefs about government and actual experience you might have with public service. Specifically, we need to look deeper at the insidious ways the climate hostage situation operates and how politicization and propaganda are actually two of the main drivers of climate anxiety. This gives you an opportunity to liberate your thinking and recover a growth mindset. Then you can begin to explore the positive aspects of your political consciousness in relation to nature and climate. Getting clarity about your "politics of the Earth" is healthy, even inspiring and fun. It supports your personal sustainability and environmental identity and creates a pathway to action.

To stand for your environmental values, you don't need to be a certain kind of political person. You may not think deeply about your political beliefs, or you may be highly engaged in environmental advocacy and activism—or something in between. Larysa didn't consider herself a political person. She only became interested in climate politics out of concern for her family. Christina had been active in her local government for years and regularly met with her representatives as part of the Citizens' Climate Lobby. As I helped Agnes and other young climate activists in her group learn, there is an inspiring diversity of ideologies and strategies to choose from when it comes to finding your own style of eco and climate politics—more than you might imagine. You have options for how you claim your rights as a citizen of the planet.

Basic Facts About Climate Change and Psychology

Over the last 200 years, people have anticipated, theorized, measured, and documented the climate reality we face. For the last 75 years, the scientific consensus has become increasingly clear—a preponderance of evidence reveals a clear causal relationship between human action, CO_2 emissions, rising global temperatures, and resulting climate and weather changes.[4] Unfortunately, this has implications for geopolitical power and corporate profit-making. So the truth was—to quote former US vice president Al Gore, in the understatement of the century— "inconvenient." This led to the suppression of knowledge, sowing of doubt, and policy of delay that bring us to the present day.[5]

Unfortunately, three generations have now been influenced by climate change disinformation. In fact, humans are well equipped to think long-term, understand the universe, and work out complex problems in creative and innovative ways. You can see this in the seven-generations thinking of the Haudenosaunee and other Indigenous nations, in the observatories at ancient sites like Nabta Playa and Stonehenge, in our understanding of black holes, and in our collective responses to wicked problems like ozone depletion, acid rain, and the COVID-19 pandemic.[6] The very fact that humans have come to understand the many periods of climatic changes in Earth's history and how this era is unique gives the lie to the contention that we are somehow limited in capabilities. This book describes many ways that the psychology of individual well-being and action can help us cope and solve climate and environmental problems, but these are not *inherently* psychological problems. They are political and economic problems, issues of power and oppression. That is the real inconvenient truth.

THE REALITY OF NON-CARE FOR NATURE

When I sit with a mother like Larysa, a young person like Agnes, or a political advocate like Christina and attempt to provide some care and guidance, I know that underlying the eco-anxiety and the climate hostage phenomenon are subtle and treacherous forces. As we've discussed, part of the problem stems from how we as a society communicate environmental issues: namely, that the language of climate science tends to ignore the emotional impacts of its troubling findings and predictions. This problem is mainly solvable by these entities recognizing that they have a social responsibility to acknowledge the mental health impacts of their pronouncements.

A deeper and more fundamental problem, one most people don't fully understand or appreciate, is that caring for Earth—or taking a proactive approach to preventing harm to Earth—is simply not the task of the political and economic systems that govern our world. Outside of a handful of examples, the constitutions or founding documents of the world's nations are based on the pursuit of human progress through economic prosperity—and the exploitation, not protection, of nature.[7] Only a handful of countries, and just a few states in the United States, ensure their citizens a right to a healthy and clean environment (and none grants an explicit right to protection from climate disasters).[8] Even if any eco-rights are enumerated, protection of these is deeply contested, as illustrated in the numerous climate trials in progress around the world (such as the case lodged against the state government of Montana for prohibiting any climate change information from being factored into state energy planning).[9]

Moreover, most large businesses and corporations, unless they are chartered as some form of benefit corporation, are legally bound to pursue profits for their shareholders—not to serve social or environmental good. Any benefits to nature or human health (despite what branding or marketing would imply) are secondary to profit. In addition, environmental harms created by the business or its products are purposely ignored (so-called *negative externalities*, as we will discuss later).

For most governments and economic institutions in the world, we must contend with what my psychoanalytic colleague Sally Weintrobe has called the "culture of uncare."[10] Nature always comes second. This creates the context whereby the climate hostage process begins. The truth is, we should all assume nature is at risk, unless we personally work to make it otherwise.

A SHORT HISTORY OF CLIMATE PROPAGANDA

If you are under age sixty, for as long you've been alive the problem of human-caused global warming has been known at the highest levels of the US government. In 1965, the year I was born, President Lyndon Johnson was warned by the leading climate scientists of the day that burning fossil fuels was harmful to the planet. Through the 1970s, ensuring clean air and water and protecting species and natural places were a bipartisan political value and a general cultural norm. And some of the most important environmental regulations in the United States were supported and passed by Republican leaders. It wasn't until the 1980s, during the Reagan administration, that the *partisan split* about environmental issues and regulations began to take hold. But these divisions, as Riley Dunlap's research has documented, are largely manufactured.[11] In fact, the confusion you witness in society, the intense polarization that prevents rational and fact-based discussions, and the stifling of action on the part of governments all stem from purposeful, unscrupulous actions on the part of *bad actors*.

Since the 1970s, the oil and gas industry has engaged in spreading disinformation meant to discredit climate science. Mimicking the same process used to stifle action about the health threats from tobacco smoking, leaded gasoline, and acid rain, these bad actors employ sleight of hand to confuse the public, deflect responsibility, and downplay the severity of the climate crisis.

Back in October 2013, an official-looking packet of information came to my mailbox at Lewis and Clark College. It included a free textbook with a letter that began: "Dear Professor of Environmental Studies: The Heartland Institute asked me to send you the enclosed copy of *The Mad, Mad, Mad World of Climatism* with my strong endorsement...Also enclosed is a copy of the excellent 10-minute DVD entitled 'Unstoppable Solar Cycles: The Real Story of Greenland.'" The letter recommended the work of the "Nongovernmental International Panel on Climate Change" and was signed by S. Fred Singer, professor emeritus at the University of Virginia and a senior fellow of the Heartland Institute.

The *Mad, Mad...* book was a cartoonish satire on climate science, and the DVD was a well-produced but false depiction of human-caused climate disruptions as a part of Earth's natural cycles. Upon further investigation, I learned the late Fred Singer was a World War II–era physicist and free market advocate known for downplaying the dangers of acid rain in the 1980s, the Heartland Institute was a fossil fuel industry–funded think tank that opposed government regulation of greenhouse gas emissions, and the Nongovernmental International Panel on Climate Change is a front group made up of fossil fuel industry insiders.[12] These Heartland materials might appear legitimate to someone unversed in climate science. But they are not.

This is an example of climate propaganda. And packets like this have been sent to hundreds of thousands of teachers in the United States, from grade schools to graduate schools.

The Basic Tropes of Climate Propaganda

- Creating a false sense of uncertainty and controversy regarding reliable and established facts of climate science
- Framing deadly climate disasters and environmental health threats as future risks rather than present-day emergencies
- Shifting responsibility for the problem from profit-seeking organizations to individuals, with a "we are all to blame" narrative (for example, emphasis on personal carbon footprints over corporate and state actions)
- Rationalizing continued reliance on fossil fuels as reasonable and inevitable, despite the availability of many alternatives

---

HOW CLIMATE PROPAGANDA
HELPS DRIVE CLIMATE ANXIETY

There are many social forces that provide the backdrop for climate anxiety: lack of rights to a healthy environment, lack of guidance on the emotional impacts of environmental issues, and *outrage fatigue* from polarized politics, online news, and social media algorithms that stoke continual surprise and disgust. In this troubling mix, climate propaganda is especially toxic because it denies the reality that we are experiencing and the evidence we can see right before our eyes. It is, as Agnes correctly pointed out, an insidious form of gaslighting.

Not only does disinformation cause us to question everything we see and feel, it makes it hard to talk about those things openly. If that were not enough, climate propaganda is designed to take focus off the national and economic policies that drive climate breakdown, and instead lay the blame on us, the public, and our individual behaviors. Ultimately, this creates the perfect conditions to turn us all into climate hostages: frustrated by governmental inaction, gaslit by disinformation, and held captive by the guilt and *moral injury* that come from feeling simultaneously betrayed by our leaders and complicit in the planet's problems by the mere virtue of our existence.

COMBATING THE MYTH OF "WE ARE ALL TO BLAME"

One of the first ways to deal with climate anxiety is to take it outside of yourself and see it in context. From this vantage point, your distress is a downward spiral initiated by disinformation and political gridlock, accelerated by negative media images and manufactured outrage, and exacerbated by guilt and self-recrimination, creating a fixed mindset about possibilities for solutions. This is often made worse when you are isolated or lack family or community support.

Given the amount of propaganda to the contrary, we forget that support for climate action is global: that no matter where we live, we can all take action and do our part. It's also important to remember that just because you don't drive an electric vehicle or live in a carbon-neutral home doesn't mean climate change is your fault. You are not doomed to carry society's failures as your own.

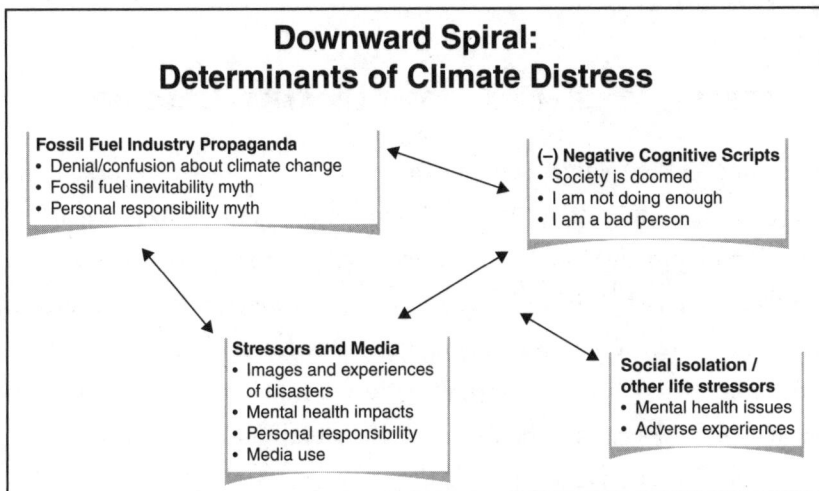

Downward Spiral: Determinants of Climate Distress

Fossil Fuel Industry Propaganda
- Denial/confusion about climate change
- Fossil fuel inevitability myth
- Personal responsibility myth

(−) Negative Cognitive Scripts
- Society is doomed
- I am not doing enough
- I am a bad person

Stressors and Media
- Images and experiences of disasters
- Mental health impacts
- Personal responsibility
- Media use

Social isolation / other life stressors
- Mental health issues
- Adverse experiences

When I became aware of how deep and long-standing the efforts were to suppress knowledge of climate change, I had to rethink my understanding of my own climate politics. Like everyone, I started with one aspect of the climate elephant I knew best: through the lens of mind and behavior, inner values and motivation, and personal action. Looking at the problem through this psychologist's lens, I naturally began helping people with their "personal carbon footprint." It was only after a deep dive into the history of climate propaganda that I could see how it had the potential to turn me and other psychologists into unwitting enablers of the "we are all to blame" myth.[13] Blaming individuals for inaction on climate issues fails to consider the structure of society: precarious economic situations, limits and barriers to making sustainable choices, and decades of false information and propaganda.[14]

The concept of an ecological footprint is a well-known metric used by economists to measure human impacts on nature. However, the idea of a "personal carbon footprint" was popularized in the early 2000s by a $200 million advertising campaign, "Beyond Petroleum," which was part of the rebranding of BP (formerly British Petroleum)—and a classic example of greenwashing.[15]

Working on your personal carbon footprint is a valid task. But focusing *only* on this allows corporations and policymakers to avoid taking responsibility for their actions.

FREEING YOUR MIND FROM CLIMATE PROPAGANDA

James, one of the young Sunrise Movement activists who worked with Agnes, was having a hard time letting go of the guilt he has internalized from climate propaganda. Together, we explored the internal messages he gave to himself (what therapists call "self-talk") when he was confronted by climate problems and accompanying guilt over his own privilege and perceived complicity. Below is a script we created that illustrates James's negative self-talk. Aspects may be true, but the overall message of self-blame is unjustified. Ask yourself if this sounds familiar.

Negative Self-Talk

1. "The world is beset with ecological problems and environmental injustices."
2. "I am personally responsible (due to my decisions or lifestyle)."
3. "My pro-environmental actions are inadequate (or nonexistent)."
4. "I feel guilty and ashamed. I am a bad person."
5. "Eco-issues are worsening. I am a failure."
6. "I try to avoid thinking about environmental problems, but each new crisis fuels a negative mood cycle."
7. "I feel despair and hopelessness."

I coached James on what a more balanced script would look like. Consider how it feels to adopt this alternate and fact-based perspective.

Balanced Self-Talk

1. "Even in the best of times, it's a challenge to govern diverse and populous societies. The system we have now, supported by governments, corporations, and powerful people, accelerates climate change and other environmental problems."
2. "I am embedded in these systems, which I did not design and have a limited ability to change."
3. "I strive to live according to my environmental values and make the best of the options I have. I give myself credit for my efforts and keep learning."
4. "Fossil fuel industry and free market propaganda promotes a 'we are all to blame' narrative about climate change that diverts responsibility from those responsible."
5. "Just because I care about climate change does not mean it is my fault."
6. "I can educate myself about climate change, act according to my values, and find satisfaction in my efforts."
7. "I can seek support and shared meaning with others who care about these issues."

8. "I don't need to solve climate change in order to cope with it."

When he compared the scripts, James became more mindful of the messages he was telling himself. To help James rewrite his negative scripts in real time, I introduced him to what I call the "What would I tell my best friend in the same situation?" technique. If your best friend engaged in this kind of negative self-talk, would you agree that they were inadequate or a bad person? Would you encourage them to feel despairing and hopeless? Or would you give them support and encouragement, reminding them of what was out of their control and the good things they were doing? Each night before bed, James recalled his day in detail and imagined what he would say to a friend who had that same day. This helped him be more gentle with himself and give himself credit for his actions.

POLITICS: IT'S BETTER THAN YOU THINK

Consider these political position statements and reflect on how closely they correspond to your understanding of current environmental problems and your own nature values (as we explored in Chapter 6):

- Wilderness has a right to exist for its own sake.
- All life-forms, from viruses to the great whales, have an inherent and equal right to existence.
- Humankind is no greater than any other form of life and has no legitimate claim to dominate Earth.
- Humankind, through overpopulation, anthropocentrism, industrialization, excessive energy consumption, resource extraction, state capitalism, father-figure hierarchies, imperialism, pollution, and natural area destruction, threatens the basic life processes of Earth.
- All human decisions should consider Earth first, humankind second.
- The only true test of morality is whether an action—individual, social, or political—benefits Earth.
- Humankind will be happier, healthier, more secure, and more

comfortable in a society that recognizes humankind's true biological nature, which is in dynamic harmony with the total biosphere.

How controversial do these statements sound? You may not agree with all of them. But they are likely reasonable to you. When these were published nearly fifty years ago in the Earth First manifesto, they were considered "radical environmentalism."[16] Now, for many they are just common sense.

Agnes was still in her teens and James was in his early twenties. Take a moment and stop to appreciate the changes that have taken place even over the course of their lifetimes: The science of climate change has become common knowledge in most schools.[17] Meanwhile, eco-anxiety and the mental health impacts of climate change have gone from a speculative idea to a basic tenet of public health and psychotherapy.[18]

It is obvious that, contrary to what we perceive as the unchanging reality of a climate hostage, how society thinks and feels about climate change is constantly changing. This includes our understanding of what the problem is, how bad it can be, and what we should and can do about it.

RECOGNIZING THAT POSITIVE CHANGE IS ALL AROUND YOU

Change is everywhere. I have witnessed people taking action and addressing climate change in homes, schools, churches, hospitals, counseling and therapy offices, governments of all sizes, businesses of all sizes, Native and Indigenous communities of all kinds, the arts, almost every branch of science, architecture and building trades, technology and engineering, farming and food production, restaurants and cooking, professional sports, law enforcement, urban and wildland firefighting and other first responders, prisons and correctional facilities (addressing heat stress on staff and incarcerated people), and the military (tracking how environmental issues impact operations and the well-being of service members).[19] The list goes on.

This vast distributed network engaged in responding to climate change is itself a benign hyperobject of sorts, too big and widely distributed to consider itself a coherent movement, more of a planetary response. It goes beyond the environmental and social justice movement that Paul Hawken evoked in his 2007 bestseller *Blessed Unrest*. It's a human species movement. It may not be as centralized, well funded, or singularly driven as anti-climate-solution forces. But certainly it is larger, healthier, has more integrity, and is more in keeping with the long-term ecological relationship humankind has with the planet over the course of human history. Arguably it will be more lasting. It is hard to know how long it will take, but the transition from fossil fuel reliance is well underway. A time when global society finally begins to arrest and roll back global heating is a high vision, but it is a vision that's possible. Many people are working on this, and you can join them.

YOUR POLITICAL VIEWS ON CLIMATE:
WHERE YOU ARE IN RELATION TO "THE SIX AMERICAS"

In 2008, researchers at Yale and George Mason Universities started an ambitious program of tracking opinions about climate change in the United States. They identified six unique audiences within the American public that responded to the issue in their own distinct way. The groups ranged from those who had the strongest belief in the existence of global warming and the greatest concern about its effects to those who denied that global warming was happening or that action was needed.

1. *Alarmed.* This group is most engaged with global warming / climate change. They are convinced that it's happening and that it's human-caused; they are very worried about its effects on them personally, and strongly support climate action.
2. *Concerned.* Members of this group are also convinced human-caused global warming is happening but see it as a more distant problem. They have less worry and motivation to take action.
3. *Cautious.* People in this group haven't made up their minds about whether global warming / climate change is happening, if it's human-caused, or how serious it is.

4. *Disengaged.* This group feels disconnected from or largely unaware of climate change.

5. *Doubtful.* Members of this group either question whether global warming is happening or don't consider it a serious risk.

6. *Dismissive.* This group does not believe that global warming is happening or is a threat. They actively oppose climate action, and they may even endorse conspiracy theories that global warming is a hoax.

Where would you place yourself?[20] Over the years, I have observed reactions that mirror climate research findings. Some people feel uncomfortable having to publicly choose one of the groups and, as in many survey situations, tend to avoid identifying on either extreme and will cluster toward the middle. Folks on the ends of the spectrum tend to be the most certain in their beliefs and consider themselves highly educated about the issue. Sometimes people get caught up in the words used to label the groups. For example, they may be convinced that climate change is an urgent problem but don't like to consider themself an "alarmed" person.

Most interesting, however, are the people who identify as "disengaged." The category is meant to describe those who are not educated or informed about climate change, or who have other priorities. (In the earliest Six Americas publications, there were pictures of people identified with the groups. The quintessential "alarmed" person was a highly educated white woman, the quintessential "dismissive" person was a highly educated white male, and the quintessential "disengaged" person was a Black single mother.) But in groups I worked with, I've found that people gravitate toward the "disengaged" category, not because they are unaware or unconcerned but because they feel disgusted, hopeless, and helpless about the degree to which climate change is—or, more accurately, isn't—being dealt with. They're climate hostages.

WORKING WITH THE SYSTEM WE HAVE

When it comes to climate politics, it's easy to feel like your actions don't matter. In a system as vast, complex, and polarized as ours, change is hard-fought and slow, and it often feels as though with every two steps forward we take at least one step back. But don't be fooled. I've learned by working with people like Christina and Marcus that working for political change matters, for two reasons.

First, the politics we have are the politics we have. There are no do-overs on government. This is not to say that political change isn't possible, but rather that climate breakdown exists right now, not in the future, and we have to work with the wildly flawed politics we have today. Globally, we are in the end game of the fossil fuel era. Oil companies and petrostates will fight hard to keep the old ways, but climate shocks will make this increasingly untenable. Weather disasters will become more severe. Resource conflicts will erupt. Rapid turnover in government will continue since no leader or party will be able to provide all the solutions needed. A desire for short-term safety and stability will continue to create pulls toward nationalism and authoritarianism, placing stresses on democracy. Fears of scarcity and displaced climate refugees will lead to economic protectionism and hardening borders. These are the political realities we are up against.

Second, the politics we have can work, even if it doesn't always seem that way. Climate change is not rooted in a fatal psychological flaw or deficiency of the human species. It is a problem that has risen from human action and decisions. Progress is and continues to be made through a mix of decarbonizing society's energy sources, assigning a price to carbon pollution so that market forces bring down emissions, geoengineering, disaster adaptation at the community level, and financial subsidies and bailouts for struggling regions and countries. Actions will be subjected to political "spin" in terms of how they are described, who takes credit, who is blamed, and who gets to tell the stories. While you can't control all of this, you can help shape those stories through your actions, your voice, and your vote. As the late American anthropologist David Graeber wrote: "The ultimate, hidden truth of the world is that it is something that we make, and could just as easily make differently."[21]

To escape feeling like a climate hostage, begin to consider yourself an engaged and informed citizen. Remember, you don't need to be a certain kind of political person. Most importantly, claiming your eco-political life requires courage and a capacity to bear witness to problems while not becoming so jaded as to turn away from life, give up on finding joy and happiness, or share your unique contribution to the world. Whatever path you choose, this much is clear: The stakes are high. We need to change what we can and otherwise work with what we are given. We are each called to the task in our own way.

Chapter 10

Anxiety

Fears of wholesale environmental destruction are of precisely this kind
—states of terror combined with utter helplessness and dependency.
— SHIERRY WEBER NICHOLSON[1]

COPING WITH CLIMATE ANXIETY REQUIRES THE ABILITY TO live with uncertainty. There is an old saying, "Don't bring me a problem unless you have a solution." We have a little voice in our heads telling us this as well. But it's okay to confront a problem without having a solution. You don't have to solve climate change in order to cope with it. In the previous chapter, we focused on how political oppression and propaganda fuel eco-anxiety. In this chapter, I'll help you understand how this eco-anxiety manifests and explore your options for coping, healing, and growth. As we move forward, keep in mind there are seldom clear-cut divisions between your experience of eco-anxiety and despair, or other sources of stress or apprehension in your life. But from a systemic perspective, learning skills to alleviate one facet of eco-distress will benefit other aspects of your mental health too.

IS CLIMATE ANXIETY NORMAL?

Over the years, many patients have asked me, "Dr. Doherty, is eco-anxiety a *disorder*?" This is a valid question. They want to know: "Is this a real issue, and if so, how big?" (I discuss how to talk to a therapist about eco-anxiety in the appendix.) But underneath this question lies a deeper set of questions: "What is normal? Are my feelings normal? Am I normal?" and even "What does it feel like when I need to question my normality?" Let's start there. Are you normal for having anxious feelings about the reality of runaway global heating, unpredictable climate-related disasters, and planetary breakdown?

In short, yes. When you are confronted with large, complex problems that defy easy solutions, anxiety is not just normal; it is also an evolution-arily useful emotional response. Arguably most sentient beings, and cer-tainly all mammals, are endowed with the ability to feel anxious about potential threats. Fear is what you feel when you perceive *certain* danger. Anxiety is what you feel when you perceive *possible* danger. Anxiety is fear wrapped in a cloud. It is uncomfortable, and for good reason. As we say in therapy, anxiety's job is not to keep you happy but to keep you alive. Like a warning light on a car's dashboard or a watchdog's bark, anxiety is a simple emotional signal to peek out the window or take a look under the hood. Unfortunately, with the meta-stresses of the modern world, we have many competing alarm signals on the dashboard of our consciousness.

HOW CLIMATE ANXIETY SHOWS UP

Consciously or not, most of us have developed strategies for coping with our eco-anxiety. But often, these coping strategies are merely our eco-distress in disguise; they are our anxiety about the capital-*I* issues expressing itself in a range of little-*i* issues. For Larysa, it took the form of constant vigi-lance and FOMO (fear of missing out) on news or information that could be important to her disaster preparedness, resulting in her endless late-night doomscrolling. Delphina's dread about forever chemicals in the Great Lakes manifested in lost sleep, unhealthy eating habits, and punishing self-criticism. Rachel was embarrassed by her anxiety about her daughter's move to college in a wildfire-prone area and insecure about her level of knowledge of climate change, so she kept her concerns secret. The good news is that each person eventually found a better recipe for healing, one that honored their environmental identity and values and allowed them to experiment with action.

THE PUBLIC AND PRIVATE FACES OF ECO-ANXIETY

Eco-anxiety is both a physical reality and a social reality. It is certainly a real and tangible experience. You are aware when it is infiltrating your thoughts, dreams, and imaginings, and we can measure the effects on your heart rate, blood pressure, and observed emotions. But we cannot reduce eco-anxiety to an individual malaise.

Chronic fear of environmental cataclysm that comes from observing the seemingly irrevocable impacts of climate change and the associated concern for one's future is part of the *structure of feelings* for the twenty-first century, and one of the defining features of our age. Eco-anxiety tells a story about society and culture, much like the collective anxiety about nuclear destruction did during the Cold War. When our emotional responses to environmental crises are not represented in scientific reports, economics, laws, and policies, anxiety bubbles up in the media and popular culture.

Eco-anxiety is also a form of social identity that signals that you are awake and care about nature and ecology. To express eco-anxiety is to become part of a worldwide community that shares your feelings across cultures, classes, identities, and languages. It can be voiced as a form of self-expression, as protest, to raise an alarm, or to engage the public in action. In each case, the process of going from suffering in silence to speaking out about your eco-anxieties and fears can be liberating.

Unfortunately, it's difficult to have a simple conversation about eco-anxiety. Like other political issues, climate disruption is spoken about in a polarized way that is itself "polluted" by propaganda and disinformation, as we have seen.[2] Being open about eco-anxiety comes with the risk of dismissal, disapproval, mockery, or even physical danger depending on your audience. Even people united in feelings of anxiety and concern can argue bitterly about the best strategies to express these feelings in order to promote societal action and avoid perceptions of doomism. But the validity of your feelings is not solely reducible to their utility in a movement.[3]

ECO-ANXIETY AND YOUR BRAIN

Your brain is a network of billions of neurons, firing off day and night, able to configure itself into patterns of endless complexity, allowing your mind to have quick instinctual responses as well as slower, more thoughtful processing. The latest neuroscience sees the brain as an organ for predicting and budgeting your body's energy needs so that you can efficiently maintain your metabolism, keep your physical systems in balance, and survive. When your brain is trying to learn something new that it expects will be important for your future survival, you have a massive increase in physiological arousal that manifests as feelings of anxiety.[4] Trying to make sense of environmental crises and the vulnerabilities they bring is incredibly taxing, both intellectually and emotionally. Simply put, it requires a lot of energy. It's no wonder people might long for simple answers or avoid thinking about ecological issues altogether.

To do its job, your brain must also be a pattern-seeking machine, constantly scanning the environment, parsing sensory data, predicting based on past experiences, and checking up on its predictions to make sure everything makes sense. Through this process, the brain is constantly sculpting its own wiring. If, like Larysa, your brain has learned to interpret information about past or future climate dangers as a current harm (activating her fight-or-flight response and her disaster-prepping tendencies), this needs to be unlearned. You can't just *tell* your amygdala to change; you need to *teach* it through experience.

With Larysa, we set up a series of learning practices where she could purposely expose herself to threatening disaster anxiety triggers (through imagination, talking, journaling, viewing news, or attending public events) while simultaneously calling forth a relaxation response from her body. She would remain mindful, slowing her breathing, keeping her muscles relaxed, and giving herself reassuring messages. Through this process new connections were made and her overactive fight-or-flight response was brought more under her control.[5] She could still feel her anxiety and it could do its job of alerting her to potential threats, but it no longer hijacked her body and mind.

WAITING FOR THE OTHER SHOE TO DROP

If your threat detection system has been primed like Larysa's, you might be familiar with the sensation of "waiting for the other shoe to drop"—that is, holding your breath in anticipation of some upcoming catastrophe. This is a common sensation and it comes up a lot in eco-anxiety. But it's something to question. Some years ago, not long after losing my wife to breast cancer, my therapist suggested to me: "Thomas, the shoe has already dropped." She was right. I was still bracing myself for tragedy long after the reality of the tragedy had occurred. And this is in many ways the case with climate change. If you can let go of the idea of climate change as a shoe that might drop in the future and accept that it has arrived—it is here, right now, and in this minute—this may allow you to reset your nervous system.

Eco-Anxiety and Trauma

In extreme situations, eco-anxiety can be accompanied by actual injuries or intense trauma brought on by life-threatening situations. These experiences can lead to a dramatic rewiring of the brain, resulting in post-traumatic stress disorder. These situations require a higher level of support than I provide in this book and are best addressed with help from loved ones and medical or psychological professionals.

LIVING THE QUESTIONS

When examining your eco-anxiety and trying to make sense of your situation, it is often necessary to hold a space for multiple perspectives, opposing ideas, and contradictions that are not easily resolved. In therapy, this is called *dialectical thinking*. Rather than answering the questions, your task is to *live the questions*—and then to learn from the process of finding answers.[6] Coping with eco-anxiety challenges you to live some of these questions:

- *Opening up versus containing.* "Do I want to let loose with all of my concerns or keep them under wraps?"
- *Being versus doing.* "Can I accept myself as I am, or must I be doing and changing?"
- *Self versus world.* "What are the boundaries between myself, my body, my identity, and the larger world?"
- *Local versus global.* "How can I be grounded in my home place and also operate within larger scales of Earth's ecology and its systems?"
- *Inside versus outside.* "Are climate problems inside me (caused by my feelings and actions) or driven by unsustainable systems outside of my control?"
- *Mitigation versus adaptation.* "Should I (and we) prioritize reducing global heating or focus on adjusting to a climate-changed world, along with all its threats and opportunities?"
- *Adjustment versus liberation.* "Must I learn to survive and thrive in society as it is, or work to change society and its structures?"

There are no right answers here. How you approach these dilemmas will depend on your situation, needs, resources, abilities, and core values, and on the opportunities thrown your way. You may toggle back and forth on these dimensions multiple times a day. What that tells me is not that you are confused but that you are in the game: alive, conscious, and actively seeking balance.

PUTTING THERAPY TO WORK ON YOUR ECO-ANXIETY

Ecotherapy refers to counseling and mental health practices that include your experience of the environment and relationship with the natural world. This could include taking psychotherapy meetings outdoors or bringing your concerns about environmental threats into the office or clinic. There are hundreds of different approaches and techniques for counseling and psychotherapy, focusing on your mind, your body, your feelings, your behavior, your relationships, your family, your gender, your culture, your age, your spirituality, and other attributes in a variety of combinations.[7] There's no one-size-fits-all approach. The best outcomes for eco and climate therapy, as with any therapy and counseling, usually depend on the quality of the relationship between the client and therapist and the fit between what the client is seeking and what the approach provides. However, the techniques below are also effective outside a therapy context. You can try them with your therapist, but you can also do them at home, without professional guidance.

Finding Your Choice Points

Acceptance and commitment therapy (ACT) helps people accept their thoughts, feelings, and situation and commit to actions based on their values. One hallmark of ACT is the idea of a *choice point*, a moment of time where you have the opportunity to be conscious of how you may be emotionally hooked by a situation, so that you can get unhooked and move in a healthier direction.[8] (This is similar to the Hook, an exercise Marcus used, described in Chapter 3.) When Larysa reflected on her climate anxiety, she began to see many choice points in her day: in her work, in her parenting, in her inner thoughts, in her purchases, in her attention to news and events. Each moment was an opportunity to stop, be mindful, and move toward her personal sustainability rather than become hooked by anxiety and confusion. She became more proactive about coping, keeping in mind her well-being pyramid, media plan, and her goal of maintaining a growth mindset. Now these concepts began to work together in a positive synergy. Larysa refocused her fear about disaster risks toward the issues most relevant for her and her family. She could say, "I've begun to learn about the risks to my home and family, and about all the services in my area that are working to protect me. I have created some basic disaster plans. There is more to do. But, for now, I know." She went from feeling powerless to feeling in control.

Recognizing Generational Trauma

Often, it will emerge that a client's eco-anxiety is related to previous traumas in their life or family. Larysa had more reason than most to believe in the real possibility of present-day disasters. She is of Ukrainian ancestry, and her childhood memories included family stories about the 1986 Chernobyl nuclear reactor accident that spread radioactive contaminants across the former Soviet Union and Europe. She still had aunts, uncles, and cousins who spoke vividly of the disaster and its aftermath: the evacuations, fears of poisoned soil and water. As a result of this history, her own parents had always been anxious about their children's safety—a style they clearly had passed down to her. Acknowledging generational trauma we are carrying helps us better understand our own anxiety and have more compassion for ourselves and our ancestors.

Making Sense of Climate Dreams and Nightmares

Some helpful techniques for healing eco-anxiety are timeless, like working with your dreams. A turning point in my meetings with Roman, the glaciologist, occurred after he shared a terrifying nightmare he'd had after his return. It was an ominous montage of images of the melting glaciers of Greenland, a dreary landscape of beaches piled high with bleached animal skeletons and whale bones, of soot-faced early Arctic explorers huddled around oily fires, and an image of city-sized glaciers collapsing, like the one captured in 2008 by filmmaker James Balog in his documentary *Chasing Ice*.[9] At one point in the dream, Roman found himself and his fiancée, Sandra, crawling their way across shifting ice floors. Sandra yelled out and then disappeared from view, and Roman feared she had fallen through the ice. He searched for her frantically and then was jolted awake, relieved to find her sleeping at his side.

There was much to unpack in the dream: Roman's dark fascination with the planet's deteriorating ice sheets, the weight of his scientific knowledge bearing down on him, his surprise at the intense love and desire to protect Sandra that he felt in the dream. In the short term, I coached Roman to go back into the dream and use creative and even fantastical dream imagery to rewrite the ending—in other words, use the dream to solve the dream.

Roman experimented with a number of creative endings for the dream, even imagining some benevolent hooded seals who rescued Sandra from the water. Finally, he reimagined that he and Sandra were wearing protective gear and were attached by a safety line. He was able to pull her to safety and find their way back to base camp.

Finding Compassion and Setting Boundaries

Delphina's research on "forever chemicals" in the Great Lakes had filled her with dread. She channeled her worry about the future into her work, where it showed up as anxious perfectionism and self-criticism ("I'm not good enough," "I should be better," "I need to work harder"). When she began to question her desire to have children, her anxiety focused on how she would bring this topic up with her family-oriented Greek parents, fearing that she would be letting her family down. She felt a constant heaviness in her chest. She lost sleep.

Delphina and I used techniques from compassion-focused therapy that help those who struggle with shame and self-criticism. Delphina recognized that activists like herself needed to be especially careful to tend to self-care and compassion because of the high risk of burnout.[10] She practiced feeling compassion for her own suffering and the suffering of others (in her case, this included other animals), and she made a commitment to prevent and relieve suffering whenever possible. Delphina also found compassion helpful when balancing her own wishes with those of her parents, and began the process of opening up to her family about her reluctance to bring a child into the world.

Drawing on the work of therapist Nedra Glover Tawwab, I helped her adapt some principles for speaking and setting boundaries with her family. These included:

1. Decide what a successful relationship will look like.
2. Ask yourself: "What can I control?"
3. Increase your tolerance for difficult conversations. Know that family members will take things personally.
4. Find a healthy distance. Take your time, and address issues when you feel ready.

Becoming Conscious of Early Life Influences

My client Rachel was also a psychotherapist, and together we adapted *psychodynamic therapy* to help her uncover the influence of her early life experiences on her present anxiety about her daughter, April, choosing to attend college in a region where wildfires could threaten her safety. Creating a safe space for Rachel to express her self-doubts and unwanted feelings helped her come to see that her reality-based fears for April were also activating her own buried hurt and anger over her own mother's negligent parenting habits, as well as her own lingering need for safety.

Rachel worked to enlarge her comfort zone as a parent and to trust April's own wisdom as a young woman. She came to see that April was better off than she herself had been in that April had *both* a sense of independence *and* a safe relationship with her mother (in psychodynamic terms, a secure base to return to after exploring). I observed that Rachel was the one who had provided all this for her daughter. This helped her see that she was a *good-enough climate parent*—the best we can hope to be—even as she was still learning and growing.

A simple way to gain insight into some of your own unconscious associations or early family influences on eco-anxiety is to start with *the private*. What parts of your eco-anxiety or distress about climate issues do you keep to yourself and not share with anyone? What do you think motivates you to keep those a secret? Are there experiences or lessons you learned early in your life that are influencing you to keep your concerns to yourself or to act in other ways? Do you see parallels to other parts of your life? Notice how you typically defend yourself against anxious feelings: Do you use humor, intellectualize, channel your angst into action, or act out and look for someone to blame and punish?

The key is to be as conscious as you can about your little-*i* issues and how they might relate to your capital-*I* climate concerns.

HONORING YOUR FEELINGS AND ENERGY

The goal of therapy is not to make you into a new person, but rather to help you become a better version of yourself. With eco or climate therapy, this often means finding new ways to be your best self on the planet that is, ultimately, your home.

American psychologist James Hillman provocatively titled his 1993 book *We've Had a Hundred Years of Psychotherapy and the World Is Getting Worse*. At the heart of his argument was the question of why therapy focuses on helping people process and adjust to their problems rather than waking them up and propelling them toward taking action to solve those problems. In the age of climate denial and disinformation, this is a valid question. But using therapy to process and cope with eco-anxiety or distress is not about adjusting to an unjust situation, neutralizing your anger and indignation, or "bright-siding" (focusing on the positive and ignoring the negative). It is about building the capacity to hold troubling thoughts and feelings in mind while also figuring out what to do about them. Adjustment and taking action are not mutually exclusive, just as climate mitigation and climate adaptation are not mutually exclusive. In fact, gaining confidence about your ability to adapt frees up energy for more proactive mitigation, in a virtuous cycle. You can find relief without denial.

You can think of ecotherapy as a form of performance psychology.[11] Ecotherapy makes us better, stronger, faster, and more focused, just as performance psychology does for athletes, artists, and leaders. Ecotherapy is also a form of self-care. Whether you are dealing with eco-anxiety or any other mental health issue, putting your needs aside should not be a default. The better care you take of yourself, the better you can care for the world.

EVEN AS...

Think of the notion of a "climate emotions party" I shared in Chapter 2. That example showed that even when you legitimately feel and behave one way, you can feel and act in other ways as well. Remember the point is not to get rid of your anxious feelings but to access other, more positive ones. Your eco-anxiety will cease to be the center of attention once it begins to mingle with other feelings like love, care, courage, and hope.

Tell yourself: "Even as I am feeling, doing, or experiencing _____, I am also feeling, doing, or experiencing _____." For example:

- "Even as I am feeling anxious about my child's future, I am striving to have gratitude for this beautiful day and to enjoy the present moment."
- "Even as I acknowledge barriers and dangers, I also acknowledge positive efforts and solutions."
- "Even as I confront the reality of extinction and my own mortality, I give my energies to life."
- Or my personal favorite: "Even as I feel eco-anxiety, I know I have the ability to navigate through it and even ride its energy toward new knowledge and possibilities."

Chapter 11

Despair

Joy, too, can reveal the world as lover and a self,
but pain is the most patient and powerful of teachers.
—Joanna Macy, *World as Lover, World as Self*

In his memoir *Warmth*, climate movement organizer Daniel Sherrell observes that sadness offers "a certain lucidity" in the face of climate break-down, suggesting that a person who is depressed is "the one best prepared for the apocalypse, the one who could see it most clearly."[1] Sherrell even comes to treasure moments when the depth of his grief seems to match the gravity of the "Problem," as he calls it. In this chapter I'll help you take a direct and clear-eyed approach to tending your despair about eco and climate issues: managing grief related to the loss of nature and overcoming depression about the state of the world. As with anxiety, grief and depression about the state of the natural world can range from normal and healthy feelings to a debilitating impairment. Given the current state of the world, it would be highly unusual *not* to feel down, sad, angry, irritated, or hopeless at times. If you have strong eco-values, the violation of Earth feels personal. When nature is assaulted, you hurt too. Like Sherrell, I recommend that you consider your despair as a way of seeing clearly—even as a creative process. To contend seriously with the problem of climate change and ecological breakdown you first have to let it in and sit with it, compassionately and honestly, "until you can observe it without blinking."[2] Despair is a passage to move through, not a wall that stops you in your tracks.

WHAT IS ENVIRONMENTAL GRIEF?

Environmental or ecological grief is the troubled feeling about past, current, or future losses associated with climate change and other disruptions to the natural world and human life. Losses can be personal or global, tangible or symbolic. Sometimes grief about nature is a private experience; other times it is experienced collectively, as in the worldwide outpouring of shock and grief that followed the felling of the Sycamore Gap tree at Hadrian's Wall in northern England, or the catastrophic wildfires in Los Angeles County.[3] Given the social pressures to keep silent about environmental threats, most eco-losses are experienced as *disenfranchised grief*—the uncanny gulf or absence we feel as lost beings and places go unrecognized and unmourned.[4]

In order to heal from your ecological grief, you must empower it by giving it a voice and standing.

When in the mid-2000s British psychotherapist Rosemary Randall developed the Climate Conversations program, one of the first climate change support groups, she looked to the findings of grief researcher William Worden and found that the stages of mourning Worden had outlined neatly mapped onto the process of ecological grief. These were not the familiar stages of denial, anger, bargaining, and so on. Instead, Worden viewed grieving as a series of tasks that can be embraced, refused, tackled, or abandoned:

1. Accepting the reality of the loss (intellectually and then emotionally)
2. Working through painful feelings about that loss (despair, fear, guilt, anger, shame, sadness, yearning, and disorganization)
3. Adjusting to the new emotional reality
4. Reinvesting emotional energy into life

Mourning is always difficult. It requires simultaneously letting go of the world that once was and embracing the world that now is. Worden also identified blocks and detours that show up in the process: struggling to accept the loss, shutting off emotions, a fixation on idealizing what is lost to avoid the pain of letting it go, feeling permanently disorganized and unable to recover, and giving up on life and love.

As you consider your own experience of eco-grief, remember that we grieve in character. Depending on your temperament, your grief may be stoic or expressive, disorganized or highly controlled, quick to manifest in feelings or slow and measured. The grief process is also fluid.[5] People have different paces and may work on multiple tasks simultaneously. The ultimate goal of grief work depends on the person and the loss. Most times it is best to think of a successful mourning process not as some kind of neat or permanent letting go but as one that helps you best manage your *continuing bonds* with what you have lost.[6] This can mean integrating the memories, documenting the history, acting in the spirit of what you have lost, or continuing a tradition in new ways. Healing from grief is like recovering from a physical wound: Your life may be different after the loss, but you can learn to live more fully within your new reality.

CREATING A GRIEF MAP

The way I have been able to orient myself amid the simultaneous, multi-pronged feelings of eco-loss is by creating a *grief map* that allows me to track my emotional response to past, present, and future losses.

You can use a map to differentiate between (1) feelings you have for current environmental problems, (2) losses and absences from the past you are aware of, and (3) losses you anticipate for the future based on your understanding of the pace and consequences of global environmental change.

On the left side is the loss; on the right side are possibilities for grief and mourning. You can begin by bringing to mind a loss or concern and then using the map to locate it. Is the loss in the past, the present, or the

Grief Map

Anticipatory Grief	Embrace the World That Will Be
What We Stand to Lose	"Lost Possible Self"
Damage and Loss We Can Prevent	

Future Losses

Unmitigated Climate Threats	Emotional Triage
Rapid Environmental Changes	End of Life / Hospice View
Losses / Extinctions	Solastalgia / Caring for Damaged Places

Present Losses

Past Losses

Gone But Not Forgotten	Memorial/Living in the Spirit Of

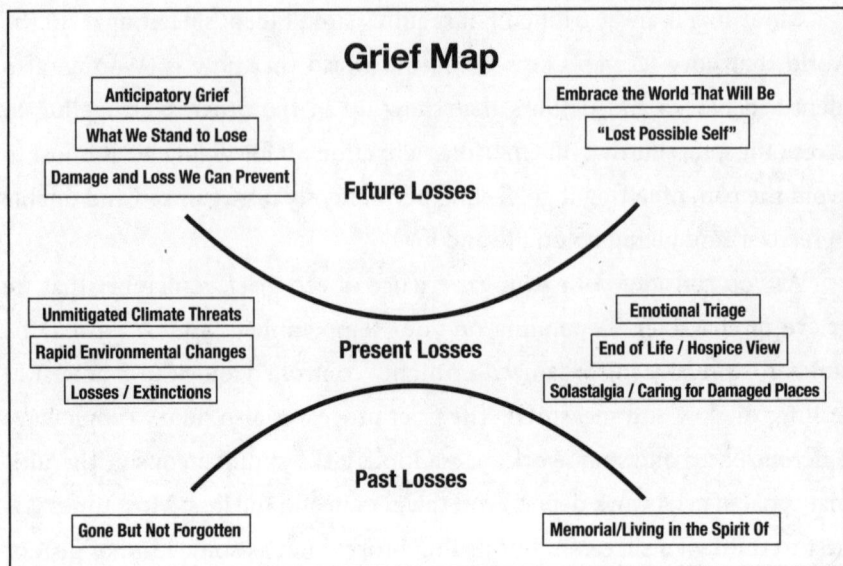

future? Each of these time dimensions deserves its own expressions for coping.

With losses from the past, we can create memorials to honor and stay bonded to the spirit of lost beings, species, and places through art, photographs, stories, and celebrating anniversaries.

For present losses, we can do what might be called *emotional triage* as we prioritize what to focus on and feel for. It can also be useful to adopt a *hospice mindset*: that is, to sit (literally or metaphorically) with damaged or dying places and species and offer what care we can.

One of the challenges of environmental grief is that it is partly *anticipatory grief*. Our knowledge of Earth's systems brings sorrow when we contemplate what we stand to lose and, more painfully, what we know we *will* lose in years to come. As you consider the losses that may occur in the future, it's important to realize that you also have some measure of control over these possibilities. Environmental grief is not just reactive. You can take proactive measures to protect things you love that are threatened.

One unappreciated dimension of climate grief is mourning a *lost possible self*, a concept that I learned from social psychologist Laura King.[7] Think of a lost possible self as the future you imagined before a life-changing event—a death, a relationship breakup, a serious injury, or a career setback—took away the possibility of that future. In the era of climate breakdown, we might mourn the loss of a future in which it is safe to be outdoors during the hottest months of summer, or one in which our homes are not at risk of being destroyed by hurricanes or wildfires.

Until just a few decades ago, we (that is, humans) took Earth's climate for granted. Now, with the stable Holocene era under threat, we are facing a lost possible self of global proportions, not only in our individual hearts but as groups, societies, and nations. The lesson from social psychology is that while focusing on a lost possible self is painful, it also brings a new perspective on your life, and with it the possibility for wisdom and deeper forms of happiness.

YOUR HORIZON OF HOPE

It's a sad commentary on the hard times we live in that hope has become suspect, viewed as a kind of "hopium," an opiate of the delusional masses. But as Jane Goodall wrote: "Hope is often misunderstood. People tend to think that it is simply passive wishful thinking…This is indeed the opposite of real hope, which requires action and engagement."[8] Or as Greta Thunberg puts it, "Once we start to act, hope is everywhere. So instead of looking for hope, look for action."[9] My favorite definition of hope is one I learned from Portland environmental leaders Dick and Jeanne Roy: "Hope is one's highest vision of the possible."[10] Their approach appeals to my pragmatic style. It gives me a process: sort through my visions of the future, choose the one that is most desirable and realistic, and commit to it. Of the many varieties of hope, I choose the hope that has a chance of getting done.

By seeking your highest vision of the possible, you approach what I call your *horizon of hope*. This horizon of what is possible is not static. It will naturally contract after a loss or a setback, as your nervous system draws in, like a sea anemone or a tortoise in a visceral survival response. During a climate disaster, your horizon of hope can shrink further, to the point where you can see nothing beyond surviving in the present moment. The good news is that as you recover, your horizon of hope naturally expands. A key aspect of *personal sustainability* is riding these cycles of hope. When you are at your most hopeless is when you must summon your ability to tend to eco-despair.

THE COGNITIVE TRIAD

The cognitive triad popularized by psychiatrist Aaron Beck views depression as a confluence of three factors: negative views about the world, negative views about yourself, and negative views about the future. We all experience each of these at times in our lives, and individually they are generally manageable. You can cope with negative thoughts about your future by focusing on doing something positive in the present moment. When you feel bad about yourself you can take inspiration from others. If you are unsatisfied with the present, you can envision better outcomes. Beck's insight was that when you experience all three negative views simultaneously, it is a recipe for depression. There's no obvious direction for your energy. Your horizon of hope shrinks, possibly to nothingness.

One benefit of the triad model is that it gives us three options for where to begin our coping. When James was feeling bad about himself, he could use the "What would I tell a best friend?" technique we discussed in Chapter 9. When Larysa was feeling despondent about the state of the world, she could moderate her apprehensions by changing her habit of late-night doomscrolling and increasing her sense of present-moment safety. When Rick, a seventy-year-old climate lobbyist and colleague of Christina's, was despairing about the future, I reminded him that he was not dead yet (using some dark humor referencing an old Monty Python skit he was familiar with). Neither was his potential for making positive changes on the planet.

When you find some light in one part of the depressive triad, it shines a little light on the others as well. This is how a downward spiral can be transformed into an upward spiral of possibility and hope.

OVERCOMING ECO-DEPRESSION

With climate change and other environmental problems, our ability for empathy, to take on the world's pain in our own body and mind, is a sign of a healthy connection between the systems of life of which we are a part. *Empathic distress* occurs when we become flooded with this pain and feel incapable of bringing positive change. In brain scans, empathic distress shows up much like chronic pain, depleting our dopamine levels and blunting our capacity for pleasure and reward. Just as in situations of war and human conflict, it is natural to want to turn away and limit our empathic response to protect ourselves.

"ALL THE SIRENS IN THE WORLD": TURNING PAIN INTO COMPASSION

While suffering is limitless, our capacity for empathy is limited by our own reserves of energy and metabolism. Whereas empathic distress depletes our dopamine levels, eventually leading to feelings of burnout and depression, compassion shows up in our brain and nervous system as a caring impulse associated with release of oxytocin. When empathy depletes us, compassion has the power to feed us.

Years ago, I spent time living and working on the remote North Rim of the Grand Canyon in northern Arizona. One day, at an isolated viewpoint high over the canyon, I did a solo psychedelic journey using psilocybin mushrooms. Alone, except for my patient dog, and miles from any roads or habitation, I heard sirens. But not just sirens, I had a sense of hearing *all of the sirens in the world*: every siren from every accident, every ambulance, every disaster, every war zone. It was an overwhelming sensation of absorbing all of Earth's pain and suffering, and it lasted for what seemed like hours. Though I held this troubled memory for many years, I chalked it up to a youthful, unguided "bad trip." It was only recently that I was able to reinterpret the experience as a lesson in the power of turning empathy into compassion.*

---

* I discuss proper facilitation of psychedelic journeying in Chapter 16.

Using a similar technique I described earlier with Roman's dream, I imagined being back in the journey, hearing all of those sirens. But rather than recoiling from them, I sought them out, traveling to every location and every place of pain and suffering and offering to help in any way I could: carrying, cleaning, cooking, soothing, burying. And in visualizing myself taking these actions, the burden of all the pain I had been carrying was transformed into compassion. Now, each time I consider that siren journey, I see it as a rite of passage.

A "TIME OF ASHES": SAD SONGS AND YOUTHFUL DESPAIR

Earlier we met James, in his early twenties, who was in the same local Sunrise Movement chapter as Agnes. James also worked with various mutual aid groups in Portland handing out water to unhoused people during heat waves and exploring ways to use new technologies like blockchain to create small-scale community economies that could be more self-sufficient and resilient in the face of climate change. In some ways, James was a model of youthful despair. He had much to be thankful for. He came from a supportive, progressive family. He had been a competitive athlete and was a youth sports coach. He studied satellite imaging at his liberal arts college and had already built up substantial savings by investing in cryptocurrencies. But now, back home after graduation, a cisgender white male amid his highly diverse and socially conscious Sunrise peers, he was burdened by a massive sense of privilege and guilt. The sense of inspiration he had in college had waned. Now he was just "Mr. Man with Impossible Plans" (a reference to a song by musician Elliott Smith).

James dismissed his feelings of anxiety and despair as "first-world problems." "I know," he told me, holding up his fingers in air quotes: "stereotype of the sad white dude." I called him on this: "James, are white dudes not allowed to be sad?" While helping James see that his feelings were valid, I also coached him to be more aware of how he had been denying himself the compassion and respect he sought to give to others.

Part of my work with James was a standard depression assessment. How new were his feelings of being sad and blue? Were they ever severe enough to keep him in bed for days or weeks, or cause him to harm himself or consider suicide? When were the exceptions, when he felt happy and like a good person? James admitted to periods of sadness throughout his life and believed depression ran in his family. He had thus far been able to cope without counseling or medication. He admitted he had fantasized about what it would be like to give up his life as a protester or to protect others but never felt suicidal.

I helped James reflect on his temperament, how he shared personality traits with his parents, and how, like everyone, he had a "happiness set point"—his usual baseline mood, something to discover and learn to accept and accommodate. I also observed that in developmental terms, James was dealing with what poet Robert Bly has called the "time of ashes," a period in life when a young person must confront the reality that the world is much more flawed and unjust than they imagined. As often happens, James's sheltered life had left him more vulnerable to a rude awakening about the state of the world.

I connected with James's despair through music, specifically the old vinyl records he collected. It was a hot, smoky summer. Alone in a garage apartment at his parents' home, an old air conditioner rattling, James ruminated on the apocalyptic imagery he heard in the Doors song "The End"—children going insane, waiting for summer rain. We played a game of trading references to lyrics. "Hello darkness" answered by "my old friend." "Don't it always seem to go" answered with "you don't know what you got till it's gone." A breakthrough occurred when I reminded James that he was not alone now or in history. Many others have felt the pain of the world, referencing Paul McCartney's lyrics to "Hey Jude." When he felt pain, could James refrain from the impulse to carry the world on his shoulders? Could he let out what he was feeling and let in love and beauty, even in a damaged world? For James, the first step to healing was the recognition that he could take a sad song and make it better.

STRUGGLING WITH THE SEASONS: SUMMER SAD

When I met with Delphina, the veterinary student researching PFAS, our first priority was helping to ease what appeared to be a case of summer seasonal affective disorder (SAD). Summer SAD shares the distinct seasonal onset of sad mood and diminished pleasure of its better-known cousin, winter SAD, but rather than sleepiness and sluggishness brought on by the lack of sun and limited daylight, it typically manifests in agitation, irritability, and insomnia and is thought to be triggered by the long days, bright light, heat, and humidity. And not surprisingly, it is expected to grow worse with global heating. (See the appendix, "How to Talk to Your Therapist About Climate Change," for tips on how to seek help with issues like summer SAD.)

There were many little-*i* issues that contributed to Delphina's capital-*I* climate despair: the stress and fatigue of graduate school, her high standards, her troubling knowledge of toxins and climate issues, her apprehension about her childbearing decisions and how her family would react. In the short term, we needed to find a way to get her back into some healthy rhythm with the seasons. Symbolically, since the sun had become her nemesis, I suggested she reacquaint herself with the moon. I also meant this literally; could she practice being attentive to the moon phases and the changing patterns of animal behavior at dusk and in the night, and connect with a gentler aspect of Earth after the harsh summer sun set?

This rebalancing turned out to be helpful. Delphina realized she had become exhausted from pushing herself in the heat of the day. She adjusted her schedule to do her dissertation field work in the early morning and evenings. She started practicing a traditional siesta, and even returned to a practice of keeping a We'Moon astrological and moon phase calendar like she had back in high school.[11] I suggested she forgo electric lights some evenings and use candles instead, and plan some overnight camping trips, as sleeping outdoors without artificial light is proven to reset our circadian rhythms and be helpful for insomnia.[12]

Delphina's sleep slowly improved. She found a renewed sense of calmness and balance. Now we could begin to do more targeted therapy to focus on those little-*i* issues that left her feeling a toxic triad of guilt, powerlessness, and doom.

"WE LOST IT": ELDERS, LEGACY, AND EXISTENTIAL DESPAIR

Rick was an economist by training and had been a proponent of sustainability since the 1970s. He was deeply disillusioned about the fact that the climate crisis had worsened and even accelerated under his generation's watch. When I first met Rick he declared, "We lost it." That is, he believed the environmental movement of his generation had failed.

Rick also had other reasons to despair. Once a vigorous outdoors person and avid body surfer in ocean waves, he now suffered from COPD and shortness of breath resulting from many years of smoking. A contentious divorce had left him alienated from his former spouse and his daughter, which also made it difficult for him to see his grandson, who was studying environmental science and whom he wished to mentor. Meanwhile, Rick's current life partner, Tina, was tired of his constant complaining about environmental and political issues. While Rick found moments of peace and clarity in his long-standing practice of meditation, he feared he had become a bitter old man.

Rick felt like a climate hostage, and when he considered his legacy, he was angry and resentful. In our first session, I validated what he was going through and asked him how he coped. Like many people struggling with despair and other losses, Rick referenced Viktor Frankl, the Nazi death camp survivor whose writings celebrated a person's ability to find a sense of meaning in even the darkest situations.[13]

Rick was signaling that his climate distress had reached an existential level, to the point where it was threatening his core sense of meaning and purpose in life. It's not a coincidence that one version of ecotherapy takes its cues from existential therapy, because climate breakdown brings up the core questions about the nature of human existence: life, death, freedom, responsibility, and so on. An existential approach to eco-despair asks not "Why are you feeling hopeless or suicidal?" but rather "Why do you want to live? And what are you living for?"[14] When I am working with someone and we are going deep, at the boundary of their feelings and lived sense of being in the world, we realize that to be open to life is to be triggered. Our lives can feel captive or free, inauthentic or authentic, responsible or irresponsible. We can be in touch with the inevitability of our own death, our own privilege, our own oppression, our own anger and rage, our own shame or survivor guilt, our own acceptance and forgiveness, our own choices and actions. We realize it's our responsibility to make the most of our one precious life. Ultimately it's on us.

With people like Rick, James, and Delphina, we go after the biggest capital-*I* issues, but it's also important that they honor and find meaning in the small daily acts of their existence. To quote a Zen Buddhist ideal: "Before enlightenment, chop wood, carry water. After enlightenment, chop wood, carry water." This was the time to get back to doing work that was important to Rick—to "chop wood and carry water."

In teasing out the capital-*I* and little-*i* issues that were the sources of Rick's suffering, an initial focus became the personal sustainability actions within his control, particularly health and well-being activities that Rick could engage in despite his COPD, like getting outdoors. He also began consulting a feelings list so he could share a wider range of his feelings with his partner, Tina. Rick had a sophisticated understanding of environmental science and sustainability and a passion for sharing his knowledge, but no one to share it with. So I steered him toward the local Sunrise Movement, where he could connect with young people like Agnes and James. Could finding a new community inspire him to reinvest his energy into life? Could he discover renewed meaning and purpose in mentoring a new generation of people devoted to taking up his mission? The key here was helping Rick take an approach to activism that was *life-transcendent*—with a lasting impact, or legacy, that was larger and longer than his life. (We'll discuss the importance of self-transcendence more in Chapter 17.)

DROP THE ROPE: MINDFULNESS PRACTICES

Practicing mindfulness is another useful approach to managing eco and climate concerns, as it helps you shift between "doing" and "being" mindsets.[15]

When we are in "doing" mode, we are compulsively trying to think our way out of problems by focusing on ideas over feelings, focusing more on the past and future than the present, and accepting negative thoughts without questioning them. Left to its own devices, this mode will persist until the problem is solved or you are too exhausted to do any more. While there is a time and place for committing to a task or goal and seeing it through to completion, the doing mode can be demanding and unkind, and depletes you in ways that leave you vulnerable to depressed feelings.

Mindfulness practice is a way to cultivate your "being" mode. This mode includes conscious awareness, using all of your senses, living fully in the present moment, being open and curious about unpleasant feelings, and seeing your thoughts as just thoughts. In ACT therapy there is a saying: "Drop the rope."[16] In other words, quit playing tug-of-war with the situation. The next time you feel yourself getting bound up by negative thoughts or feelings about the environment or climate problems, ask yourself: "What would it mean to drop the rope? It would be easy to go back to the tug-of-war. It's what we're used to, and it's what our nervous system knows and expects. But what if we could stop the struggle and be present in this moment, rather than focusing only on a distant imagined goal?" James found comfort in the idea that dropping the rope was an option. It helped him realize how much time and energy he had been spending trying to push away certain feelings, which is the opposite of being open to experience. (He used this in combination with rewriting his self-talk, as we discussed regarding climate propaganda in Chapter 9.)

EXERCISING YOUR HORIZON OF HOPE

As we discussed earlier, hope comes from recognizing and acting on your highest vision of the possible. To expand your horizon of hope, first focus on what you are doing today, right now and in the next few minutes. What do you hope for in this narrow window of time? It's likely attainable. Now, ask yourself what you hope for in the next several hours or in the next week. Is it within reach? When you move out further into the future, it's easier to start doubting that your vision is achievable. This is normal. In psychological terms, you're pushing the bounds of your *locus of control* from things you can reasonably control to things that are beyond your control, which puts you in danger of catastrophizing or imagining unrealistically bad outcomes. When this happens, bring the focus back to what's right in front of you. When you do, you will understand that your highest vision of what's possible is often much more in your control than you realize.

———

Even during the best of times, being a part of nature and the cycle of life will expose you to tragic events, painful losses, and reminders of mortality. In this way, tough emotions can be teachers that help you build the maturity and compassion to sustain you on your journey to becoming a *climate survivor*.

Chapter 12

Place

The world is now dominated by an animal that doesn't think it's an animal. And the future is being imagined by an animal that doesn't want to be an animal. This matters.

—MELANIE CHALLENGER[1]

I STAND IN A CIRCLE WITH THE COUNSELING students in a clearing in a wooded campground. I ask them to close their eyes, and, without breaking the circle, turn around to face outward toward the trees around us. This symbolizes turning our senses away from the human world that dominates our attention as we begin to consider the more-than-human world. Eyes still closed, I encourage them to take in the soundscape of the place, the wind in the swaying tops of the fir and cedar trees and the occasional creaking of their wide trunks, the faraway calls of birds, the buzzing of insects, the sound of their breathing. Then to feel the air on their skin, the weight of the midday summer heat, and the faint breezes. Next, the fragrance of the woods. I invite them, when they are ready, to slowly open their eyes.

We have been learning about the Japanese practice of *shinrin-yoku* or "forest bathing" and now have a word—*komorebi* (sunlight filtering down through trees)—to describe the flickering contrasts of brightness and shadow playing around us. I encourage the students to notice their impulse to reach out and explore this beautiful place, and suggest they pause and ask permission of the place, of the land, before taking a step. In their own way, they make their requests. Then, one by one, they move outward from the circle, slowly and reverently, and begin to make contact with the forest.

THE NATURAL ALIEN

The spectrum of nature settings introduced in Chapter 7 was a helpful framework for Christina, the political advocate who felt estranged from the natural world. It helped her recognize many ways she could appreciate the presence of nature—spanning from indoor nature to wild places—in her life. But she needed more. She still felt deficient and even ashamed because of her vague disconnection from the natural world.

It's difficult to be a human animal on Earth, within a culture that largely ignores the fact that we share a habitat with other sentient beings. We are a young species relative to most of the beings with whom we share the planet; we are still figuring out how to live on Earth companionably not only with other humans but also with the rest of the more-than-human world: our animal and plant relatives, the land, the sea, the sky, the atmosphere. We talk about nature in abstract terms, but we don't really see it. Most of us cannot live off the land like our ancestors and don't even know the names of plants and animals we see every day. As Neil Evernden argued in his classic 1993 reflection on environmentalism, *The Natural Alien*, once we leave the bubble of human society we can be a stranger on our own planet. It's easy to perceive this blind spot as a failure at the core of our environmental identity, some hidden flaw or original sin. But the reality is that after several millennia of living in dense human settlements (including today's human-centered globalized cities), our species has managed to survive seemingly without a fundamental understanding of the natural world.

Our habitual ways of seeing and being in the world—getting caught up in the news, our work, our daily human dramas, and the thoughts in our own head—obscures a deeper reality. We are not separate from the living landscape around us.

Part of this stems from the way nature is generally understood—or misunderstood. Science does a good job taking the world apart, but not putting it back together. Our educational system fragments our understanding of nature into separate school subjects and disciplines: natural sciences and social sciences, arts and humanities, economics, politics, theology. An unspoken part of the curriculum is that it is taught mainly indoors, on screens, with nature itself abstract and distant.[2] (For ways to counteract these problems, see the discussion of environmental education in Chapter 8.) Through this abstract and human-centric education, we become alienated from the natural environment and from the ways other species live around us in a vast set of interconnected and interdependent systems.

This educational approach is one origin of the environmental crisis. As we feel more and more disconnected from nature, we feel less and less willing and able to protect it. This nature blindness creates a psychological distance that allows for eco-destruction and animal suffering to occur outside of our daily awareness. When knowledge of ecological damages breaks through, it increases the desire to distract ourselves from this destruction, which tragically creates further space for more profound and irreversible losses to occur, in a vicious cycle.[3]

Some years ago, I was invited to do a workshop at Sundance, a beautiful ski area and arts center in the Wasatch Mountains of Utah. It was summer. The windowless theater I was assigned to was not conducive to our group activities. So I suggested we go outside, where we could practice some outdoor meditation. I quickly scanned the nearby grounds and led the group to a grassy spot. I instructed everyone to find a place to sit or lie down and invited us all to close our eyes. As we settled in and opened our senses, we heard a whirring sound as hidden sprinklers popped up out of the ground and began dousing us. Everyone had a good laugh, and as a young therapist, I was quite embarrassed. In hindsight, I should have recognized that a flat lawn in a high desert canyon had to be human-made and irrigated, and anticipated we might get caught in the crosshairs of an automated sprinkler system. I can forgive myself. I did not know that place. But it was a teachable moment. Though I was invested in "connecting people with nature," I had my own blindness about nature to overcome.

RECOGNIZING BARRIERS TO PLACE ATTACHMENT

I reminded Christina that the messages she told herself about disconnection with nature, and the separation imposed on her from society, could be questioned—and that she was not doomed. With a growth mindset, she could learn to "acknowledge the land" on many levels, not just as a place to walk and exercise in but as a living presence she could bond with and an essential part of her interconnected self.

The boundaries between humans and nature have always been artificial. To truly heal from climate anxiety, we need to speak the true story about the natural systems of which we are a part. To do this, we need more tools like place attachment and recognition of shifting baselines.

Geographers, architects, and psychologists have long studied *place attachment*: the emotional bonds and relationships people develop with both built and natural spaces ranging from homes and cathedrals to Central Park and the Grand Canyon.[4] As clinical psychologist Susan Bodner and her team at Teachers College in New York have found, emotional attachment, long studied in close human relationships with other humans, operates in very similar ways with natural places (and with other species). When asked what special places mean to them, she found that people often speak in the language of an intimate relationship. When asked how they would feel if that place no longer existed, they described an experience akin to the loss of a loved one.[5]

There are, however, barriers to place attachment when it comes to natural spaces. Unlike homes and cathedrals, the natural world is always changing. Yet we often fail to notice it, because of our *shifting baselines*: a concept in conservation science that describes how each new generation takes the current state of the natural environment as the norm, not realizing the degree to which it has been depleted and degraded by previous generations. Where I live in the Pacific Northwest, for example, we revere the towering

Douglas firs, the stunning mountaintop vistas, and the diverse wildlife. But rarely do we stop and think about how most of the ancient trees that awed the Europeans are gone or how the salmon runs that were once a wonder of the world, an entire river teaming with millions of flashing, migrating silver fish that you could hear from miles away, are beyond memory. It's almost impossible to conjure such richness. Atlantic salmon used to swarm the rivers of New England as well.

Shifting baselines also happen on a personal scale. We lose connection with regular outdoor activities and connections with places and other species, and don't recognize this until we look back and realize we have long been missing what sustains us.

In the last few years, the physical evidence of climate breakdown has created its own barriers to going outdoors and into nature for healing and restoration. At times, these barriers are physical: dangerous heat, smoke and poor air quality, burned forests and trails, red tides. Other times the barriers are emotional, as going outdoors brings up painful feelings about environmental destruction.

The more time Christina spent outside, hiking in preparation for her trek on the Via di Francesco, the more she got in touch with her physical body, one underappreciated form of reconnecting with nature, while also becoming more in tune with the local land and weather.[6] As she read the history of the Umbria region, where she would be walking, she also became more curious about the history of where she lived—another reconnection. The boundaries she had assumed between herself and the natural world began to blur, as she was slowly learning to see nature not simply as a place but also as a set of long-standing relationships. (Later, we'll talk about how she began to weave her spiritual life and the teachings of St. Francis into her nature connections.)

DISSOLVING THE BARRIERS

How can you begin to address feelings of disconnection and issues like nature blindness? Developing your environmental identity and values, as we did in Part Two, is a good start, as is understanding the full spectrum of the natural settings in which you can partake. And, like any other relationship, getting to know your local place takes spending time together, being in the present moment, paying attention, listening, observing, and interacting.

These are all what the Canadian ecopsychologist Andy Fisher calls *recollective activities*—acts that help you remember, or relearn, "the essentially human art of revering, giving back to, and maintaining reciprocal relations with an animate natural world."[7] You can think of these activities as devices for dissolving the artificial human-nature boundary we are surrounded by. Asking permission before exploring a place (as the students did in the anecdote I shared earlier) is a recollective activity, as were Christina's pilgrimage and Jesse's relearning to connect with the burned forest.

These activities start with two simple ingredients—being outdoors and being mindful. But being mindful in an outdoor setting is more difficult than you might think. That's because of the very way our modern society continually places our mind in the foreground of time and space, with nature as a passive backdrop.

In his wonderful book *The Spell of the Sensuous*, ecologist and philosopher David Abram describes an outdoor awareness exercise that I have adapted and used to become more mindful when spending time with nature. This is how Abram describes it:

> I locate myself in a relatively open space—a low hill is particularly good, or a wide field. I relax a bit, take a few breaths, gaze around. Then I close my eyes, and let myself begin to feel the whole bulk of my past—the whole mass of events leading up to this very moment. And I call into awareness, as well, my whole future—all those projects and possibilities that lie waiting to be realized. I imagine this past and this future as two vast balloons of time, separated from each other like the bulbs of an hourglass, yet linked together at the single moment where I stand pondering them. And then, very slowly, I allow both of these immense bulbs of time to begin leaking

their substance into this minute moment between them, into the present. Slowly, imperceptibly at first, the present moment begins to grow. Nourished by the leakage from the past and the future, the present moment swells in proportion as those other dimensions shrink. Soon it is very large; and the past and future have dwindled down to mere knots on the edge of this huge expanse. At this point I let the past and the future dissolve entirely. And I open my eyes...[8]

Abram then finds himself "standing in the midst of an eternity, a vast and inexhaustible present." In the moment, "the whole world rests within itself," linear civilized time ceases to exist, and the barriers between us and the natural world evaporate.

CONFRONTING THE DAMAGE

Forging emotional attachments with nature is not without risk; it can open you up to a "world of wounds," as nature's vulnerability to climate threats may reveal fears, lost possible selves, sorrows, or grief we'd rather avoid.

Jesse, who felt grief after her family's forest property was swept by a wildfire, was having a hard time as she walked the ruined landscape of her childhood. "How can I go to the trees to heal my own hurts when the forest is literally burning?" Jesse asked. The place she had always gone to for comfort was now itself suffering.

An unfortunate reality of the era of climate change is that for many people, a sudden and painful awareness of the loss and damage of landscapes they had previously taken for granted is the *entry point* into their environmental identity and understanding of nature. This is a troubling double bind that society needs to address: the conflict between the growing desire on the part of the public to connect with nature for health and well-being and the emotional toll of these otherwise healthy activities.

But we can't wait for society. Thus, we require skills for *seeing and being with damaged places*—areas that have been depleted, polluted, trashed, torn up, or overdeveloped—however tempting it might be to ignore or avoid them.

LEARNING TO RETURN TO DAMAGED PLACES

Author and environmentalist Trebbe Johnson has created a grassroots international program, Radical Joy for Hard Times, that is dedicated to helping people connect with natural places that have been damaged through human or natural acts: another great example of a recollective activity.[9] The Rad Joy, as devotees call it, has five simple steps:

1. Go, alone or with others, to a wounded place.
2. Spend time there and share your stories about what the place means to you.
3. Get to know the place as it is now.
4. Share with others what you discovered.
5. Make a simple gift of beauty for the place out of materials the place itself offers, such as natural arrangements of wood, stones, or flowers.

One of Trebbe's observations I found particularly helpful was that although people typically avoid damaged places out of anticipatory grief or fear about what they might find there, when they actually go and confront it, their experience is often quite different. Trebbe likens this to visiting a sick friend: You might be tempted to avoid visiting for fear of what you might feel, but instead, she says, "you visit them, you bring them a gift, you talk about their life, and you share your life. You keep up the connection because that person is still who they are."

Similarly, even when a place has been completely destroyed, it still lingers in our consciousness, in our spirit, in our emotions. By going there, we revive that connection in a new way. We're saying, "I'm still willing to be with you, and I still care." This is in keeping with the concept of "continuing bonds" when dealing with grief. Rather than severing our emotional ties with a departed loved one to avoid confronting our pain, healthy grieving entails finding new ways to honor our emotional attachments with those we have lost.

I told Jesse that she was not alone in her ecological grief and that it was possible to see this pain as a doorway and not a barrier. I helped her begin to work through tasks of grief as they applied to her relationship with the burned mountainside on her family's land. Once she accepted the reality of the loss, worked through her painful feelings, and adjusted to the new reality, she began reinvesting new energy into life by attending to and caring for the damaged landscapes. This process of restoration was both ecological and personal. Jesse created a mantra for herself: "As a child I was blessed with innocence. As an adult I let go of my innocence and bear witness to what is painful and damaged in the world."

Commitment to nature means being present for damaged, wounded places and the injured or displaced beings that lived there. Opening your heart and seeing the life that is there is also a form of nature connection. By learning how to mourn the places and histories we have lost, we learn to recognize places in new ways.

Remember, the boundaries between humans and nature were always artificial. If you are *part of* nature, then caring for damaged places means *healing with* nature.

FINDING YOUR PLACE TO STAND

James, the young climate activist, was tasked to create a *land acknowledgment* about the local place and why it deserved protection for an event his Sunrise group was sponsoring. But he felt unqualified, and I understood why. As we've discussed, it is quite normal to have an *imposter syndrome* about your environmental identity, to lack a sense of belonging to the land. For James, as a person of European descent, he questioned whether he was even entitled to have an attachment to any place.

A land acknowledgment is a traditional practice in places like New Zealand, Australia, and Canada and among tribal nations in the United States, where it is a policy to open events and gatherings by acknowledging the Indigenous people of that land. In recent years, many individuals, cultural groups, and schools have adopted this custom to recognize Native people's ties to land seized by colonial powers. There is no one protocol for these acknowledgments and some question whether they are just empty performances. In my experience, people are sincere about making these acknowledgments. But like James, they may only have superficial knowledge of the local environment and human history, or lack a working sense of their own environmental identity.

To orient James, I suggested he begin with a few questions: Who or what is being acknowledged, and what is the point? Where are you starting? Where do you want to end? Are you compiling a history, and if so, are you going from past to present, or starting from the present and looking back? A proper land acknowledgment requires homework about the people, geology, and native species. How did the land form? What is the history of the ecosystem, and the plants and animals that allowed it to flourish? Who were the first peoples of that land, and their ancestors? What is the history of subsequent human migrations, conflicts, and environmental shifts? And where do you fit in? At what point in historical time do you—or your ancestors— enter the story, and what is your ancestors' lineage of place? What do you offer to your current place? How are you backing your land acknowledgment with substantive actions (for example, volunteering or donating on behalf of land conservation or species, contacting and supporting local Native organizations, or taking part in local government)?[10]

FINDING YOUR STANDING PLACE

Connection—or the lack of connection—with natural places is based on more than just knowledge or emotions. Often, lurking under a sense of disconnection from nature is the fact that most people don't feel a cultural connection to the land. They lack roots.

But to what degree are cultural roots a prerequisite for feeling a connection to a natural place? If you don't have Native heritage and may in fact be part of a colonizing people, how can you claim a connection to the land? These are truly radical questions to hold and live (*radical* means "from the roots"), and there are no easy answers.

The Maori people of New Zealand have a principle called *turanga-waewae* (too-runga-why-why), which translates as "a place to stand" or "standing place."[11] I learned of this when I had a chance to visit New Zealand to work with the Department of Conservation there. Therefore, I share this information with great respect. I am not suggesting you appropriate Māori culture, but I believe there are universal lessons to be learned from the concept of having a *place to stand*, just as there are from nature kinship practices of peoples from all parts of the globe.

To establish a *standing place* in your own life, you need to locate a local place, region, or ecosystem that you can build an attachment to and commit to protect. You begin by spending time there, making offerings to the place of your time and effort, and asking permission. Your standing place is where you choose to root your environmental identity. Then you can work on integrating, in whatever way possible, your ancestral place traditions and building new, healthy traditions.

TELLING THE STORY OF YOUR PLACE

Culture lives in stories. So as I suggested to James, before we can commit to a place, or see ourselves as a person of a place (or of places), we need to be able to tell the place's story.

For most of my life, my connection with the various places where I lived was a scattered patchwork of facts. I had done some homework, but whenever I began to experiment with making a sincere land acknowledgment that encompassed all I might know and learn about a place and its human, natural, and geologic history, it felt like I was tumbling through space and time.

When I traveled to the north of Ireland, where you can find castles with the Doherty surname, and spoke with my clan historians, the 1100s seemed not that long ago and the 1600s felt like yesterday. If you looked closely, bricks from centuries-old castles were used in the houses people live in today.

Where I live now, in Portland, Oregon, humans are relative newcomers. Though the land was inhabited centuries before the Europeans arrived in the 1840s, the salmon have been returning to the same rivers here for some 6 million years, even surviving the great floods at the end of the last ice age. We think of the tall Douglas fir trees as native. But they are immigrants too. Douglas firs branched from larches about 50 million years ago and migrated from northern New Mexico to California and the Pacific Northwest at a rate of about a mile every 5,000 years.

Portland was once just called "The Clearing," a spot on the shores of the Willamette River that was the farthest place inland to which oceangoing sailing ships of the time could travel. It famously got its name from an 1843 coin toss of a copper penny between two white settlers who shared a land claim there, Asa Lovejoy of Boston, Massachusetts, and Francis Pettygrove of Portland, Maine. Pettygrove won.

Looking for Native history leads in unexpected ways to the present and back to the past. When I met Oscar Arana, the CEO of the Native American Youth and Family Center (NAYA) in Portland, we discussed tribal history, but mainly focused on the present, where the primary strategy for the Native community's resilience was wealth building, including creating a corridor of Native American–directed housing, businesses, and parks in the gentrifying neighborhood.[12] A victory was the Return to Neerchokikoo fundraising campaign to permanently secure the old school building near the Columbia River where the NAYA Center is located.[13] When commuters pass by the site on their way to the airport, they don't realize that Neerchokikoo is a centuries-old gathering site that was referenced in the journals of Lewis and Clark.

Again and again, histories we think are solid and rooted are not. Stories of place and history are continually being rewritten and retold. Take the journey of the Willamette Meteorite (Tomanowos, "the visitor from heaven," in the local Clackamas language), a fifteen-ton iron and nickel meteorite, and the largest ever found in the United States. Tomanowos rested for thousands of years in the Willamette Valley, where it was sacred to the Clackamas people. After the Clackamas were displaced to a reservation, the sacred Tomanowos was also displaced, and eventually sold and transported east to New York City, where you can see it today displayed in the Hall of the Universe at the Museum of Natural History. And then there is another slip of time and space: It turns out that when Tomanowos fell from heaven, it actually landed in Canada. Ice age glaciers slowly moved the meteorite south to the area that is now Montana, where it rested until the great post–ice age floods carried the meteor west to Oregon.

These are the true environmental stories that you are living out: works in progress over thousands and even millions of years that will outlast your lifetime.

BLACK TO THE LAND

It is not an exaggeration to say that for Marcus, the urban planner we met earlier, getting involved in the Black food and farming movement in the United States was a renaissance for his environmental identity and connection to the land. The goal of food sovereignty (the basic right of a community to control how their food is grown and distributed) opened Marcus up to the possibilities for urban farming and homesteading, BIPOC-run farmers markets, and reclaiming African American farming and nature traditions, a part of his ancestry that had been lost to him. He found collaborators in cities and rural areas across the United States, from Brooklyn to Detroit to California.[14] He was able to claim the category "working nature" on the nature spectrum we discussed in Chapter 7. He was literally growing things.

As with other waking-up moments, Marcus quickly learned that things were worse than he imagined. The food production system was rife with discrimination. A legacy of unequal access to wealth and land ownership, like urban real estate redlining, also extended to denying farm loans to people of color. Among the marginalized groups and undocumented immigrants who worked in slaughterhouses and factory farms, he discovered inhumane working conditions, exposure to infectious diseases like avian flu, and disproportionate crime and social problems.

But there were also ways in which things were better or at least more inspiring than he imagined. Through books like *The Black West*, Marcus learned the stories of BIPOC individuals who had been a constant presence in the western United States yet were often whitewashed out of frontier history. He discovered that the nearby city of Centralia, Washington, had been founded by a Black businessman. He learned that part of the historic Biden-era Inflation Reduction Act provided $2 billion in financial assistance to more than 43,000 farmers, ranchers, and forest landowners who reported discrimination by the Department of Agriculture in farm lending prior to 2021.[15]

Marcus realized that many of the climate and disaster coping needs of residents were shared by farmers and people who worked the land in rural areas. The practical, hands-on coping skills he had developed was knowledge he could teach and share. Meanwhile, working to promote community access to healthy food helped his own personal sustainability. His health and diet improved, reducing his risk for cardiovascular disease.

INTERSECTIONS BETWEEN YOUR CULTURE AND PLACE HISTORY

When you commit to a standing place (even as an experiment), your personal and cultural story will begin to blend, or intersect, with the story of that place, to become part of your environmental identity. Marcus, a Black American with enslaved ancestors whose family migrated from the US South, found that a "diaspora environmental identity" captured some of his story. There are many ways to characterize your environmental identity. For example, as the grandson of an Irish immigrant, I identify with an "immigrant environmental identity" that helps me locate my culture in both the Old World and the New. But because of my emotional attachments to several places I have traveled to around the world, "cosmopolitan environmental identity" is also fitting. Finally, I am committed to learning about and more deeply knowing the place where I live now, so an "apprentice environmental identity" also feels authentic. See if any of the identities on this list (or some combination of them) resonate for you.

Cultural Variations on Environmental Identity
- Diaspora environmental identity (your identity features ancestral or current displacement)

- Postcolonial environmental identity (your place ancestry is affected by the forces of colonization)
- Settler-colonial environmental identity (you are the descendent of settlers)
- Indigenous or First Nations environmental identity
- Immigrant environmental identity (you or your ancestors relocated to new lands)
- Cosmopolitan environmental identity (you blend multiple forms of environmental identity)
- Apprentice environmental identity (you are committed to learning your place)
- Food and farming environmental identity
- Rewilding environmental identity (focused on restoring or creating a new ecological balance)
- Climate-adaptation-focused environmental identity
- Beauty- and ceremony-based environmental identity
- An environmental identity in progress

Deeper contemplation of nature can transform your conceptions of identity, self, and reality. As you gain the capacity to see and be with damaged places, you can find a standing place to commit to, get to know, and protect, rooting your environmental identity in that place. And you reinforce that commitment by making a land acknowledgment: a statement of understanding, kinship, and purpose, based on your environmental identity and values. All these actions help you transcend nature blindness and feel more connected to your local landscape and Earth.

Later, you'll learn that your ability to be present, rather than passive, in nature will be a superpower in times of disaster, when your attention will focus on decisive actions to keep yourself, family, and community safe. Now, let's see if the foundation you have been building can support an ethical sense of happiness and thriving.

PART FOUR

Flourishing

IT'S NOT SURPRISING THAT QUALITIES LIKE HAPPINESS AND flourishing are rarely discussed in association with climate breakdown and the environmental crisis. That's exactly why I gave the podcast I host with Panu Pihkala the provocative title *Climate Change and Happiness*. It raises a question: What does it mean to be happy or well in an era of ecological crises in a way that's honest, ethical, and responsive to these times?

That is your mission in Part Four: to actively consider what it would mean for you to be happy or to thrive in the life you have right now—with open mind, open heart, open breath, and open hands.

First we'll break down the idea of happiness and consider options for what your well-being can look like. Then we'll explore ways of finding and maintaining connections in your human relationships that honor your relationship with nature. We'll survey many ways that the arts and creative activities help you to see—and to reenvision—the world and your place within it. In fact, making art, practicing design thinking, and creating beautiful things may themselves be action for you. Finally, I'll suggest how to find and express your environmental values within organized religion, personal spirituality, and your consciousness, including psychedelic journeying. Self-transcendence—being comfortable with processes larger than yourself, and taking on goals that will outlast your life—sets you up for coping and action for the long haul.

Chapter 13

Happiness

The ecological crisis reveals fundamental problems in standard contemporary views about happiness...these things cannot anymore be measured simply by material wealth, but other more holistic and ethical measures such as planetary wellbeing are needed.

—Panu Pihkala and Thomas Doherty[1]

Sam and Liberty were in the prime of life, with two young children and full-time jobs; Liberty worked from home nearly full-time as a graphic designer, and Sam was a home builder who developed affordable and sustainable housing. They were active in their community and helped to found a parents-for-climate-change-advocacy group, one of many such nonprofit groups around the world.[2]

Life was full, blessed, and hectic. Now that their younger child, Jack, was in preschool and their older child, Serena, was in first grade, life was a steady series of calendar reminders. As Sam made lunches, Liberty examined their planning spreadsheet. Who was picking up Jack after school? Liberty had back-to-back client meetings, so Sam would do the pickup for both kids, missing his afternoon workout. What about the pitch for the Families for Climate group at the local nonprofit funders meeting that evening? Liberty would attend the meeting and bring both kids to give Sam some space. Both their quality time together and their personal passions (which, in their early pre-parenting years, took the form of Outward Bound–style climbing adventures for Liberty and low-budget international travel for Sam) were on the back burner.

From the outside, Liberty and Sam didn't look like they were having a mental health crisis. They were lucky and privileged, embracing the full spectrum of human experience, or what mindfulness researcher Jon Kabat-Zinn has called living "the full catastrophe"—a marriage, children, a household, and so on.[3] But inside they were stressed out. Like one of the vet students we met earlier, they were feeling *agotado*—exhausted, emptied out, with "no drips left"—from pushing themselves to the limit each day. Meanwhile, they were worried about the future of the planet they were leaving for their kids to inherit, frustrated and angry about the state of the world, and bickering far too often over small things. In spite of all their good fortune, they were not thriving.

LANGUISHING AND FLOURISHING

In this chapter, we are going to explore new ways to think about—and reach for—flourishing while you are coping with anxiety about climate change and other global problems. Though they didn't recognize it at the time, Sam and Liberty were *languishing*. In psychological terms, languishing is a baseline where you may function well enough and lack overt problems but aren't filled with a sense of vitality or purpose. The point is to move from languishing to *flourishing*—experiencing positive emotions and happiness, having a sense of fulfillment, and connecting with others at a deep level.[4] How can you give yourself permission to flourish, without betraying your environmental identity, ethics, and values?

ETHICAL FLOURISHING

The origin of Sam and Liberty's conflict dated all the way back to their early discussions about whether or not to become parents. Having a child was something Liberty had felt strongly that she wanted, but Sam had to be convinced. He studied environmental science in college, where he learned that population growth was one of the key contributors to an unsustainable society (other factors being increasing levels of affluence and increasingly powerful technologies) and questioned whether it was right to bring more children into the world.[5] Liberty saw childbearing (or not) as a woman's right and believed strongly that babies should not be pawns in climate politics. She was also passionate about pursuing life to the fullest—"What is the point of being alive if we can't live fully?"—which for her meant having children. Sam and Liberty began meeting with other parents in similar situations, but Sam was still struggling with his mood. He needed some rationale that would give him permission to be happy about the prospect of raising children given the state of the world. He came around to the idea that he didn't have to choose between being ethical and experiencing thriving and joy.

Ethical flourishing means seeking happiness in line with your values. Your good feelings need to make sense within your social and environmental ethics, not just your personal ego-focused needs. Ethical flourishing is not about disavowal of your knowledge and responsibility about climate breakdown, or about suppressing uncomfortable emotions. Flourishing in the face of climate anxiety means that you can appreciate good things, honestly and in the present moment, while also having an awareness that things could be much better. Moments of gratitude and satisfaction have a place alongside moments of outrage and anger.

Permission to Be Happy

If you're cynical or skeptical about evoking terms like "happiness" or "flourishing" in a world that seems to hold so much suffering and injustice, that's understandable. You can pass by this section and move on to action in Part Five. If flourishing seems impossible, you can cycle back to the earlier sections of the book that lay the groundwork of coping skills, environmental identity development, values, and self-healing. Working on these can make moments of flourishing attainable on a practical basis. You might also consider how offering the possibility of happiness during times of climate breakdown might be helpful for other people you care about.

OVERCOMING BARRIERS TO HAPPINESS

The flourishing I am talking about does not call for looking at life through rose-colored glasses. It assumes you are honestly bearing witness to many profound and troubling issues associated with rapid climate breakdown. Nevertheless, there are some common barriers to face. Any message of happiness can be suspect if you are wary of *bright-siding*—what writer Barbara Ehrenreich describes as the tendency to shift attention away from the world's problems through feel-good rhetoric and advertising (a component of climate propaganda).[6] Or you may be struck by a nagging sense of absurdity when you see a crisis and everyone around you is moving happily along as if nothing bad is happening.

Friedrich Nietzsche said, "Whoever fights with monsters should see to it that he does not become one himself. And when you stare for a long time into an abyss, the abyss stares back into you."[7] When you are fighting the monsters of climate change, it's all too easy to get stuck in a dark place. But the opposite is true as well. When you bring your focus and intention on creativity and solutions, all the possibilities stare back at you as well.

But let's not throw the baby of happiness out with the bathwater of despair. My Finnish colleague, Panu Pihkala, who teaches environmental theology, shares the Finnish word *heikos*, which means both "joy" and "sadness," to underscore the point that it *is* quite possible to simultaneously hold such conflicting, divergent emotions at the same time.

That's because your emotional experience doesn't need to be constrained by a single dimension on which you must choose between positive and negative. Remember the emotional intelligence skills we talked about earlier.

You can not only express the feelings you have but also ask yourself what feelings you want to have. You can imagine a gathering or party of emotions, so that even if you feel deeply sad and angry about climate threats, there is still room to feel surprise, joy, hope, or excitement when you hear of positive developments. Sam understood all this intellectually. It's certainly what he wished for Liberty and the kids. But he had to practice daily, using the image of an emotional compass and a feelings vocabulary list to identify just the right feeling word or sensation, speaking these out publicly, and adding them to his repertoire, until recognizing good times became more natural.

DEFINING CLIMATE HAPPINESS FOR YOURSELF

I won't presume to tell you what happiness means for you in the context of climate anxiety. But I can offer some ideas about how to think about it. Much like the climate elephant with all its viewpoints, there are many ways to imagine what it means to flourish in a world plagued by climate and eco-problems. Some are individually focused, and some are more collective or relationship-based. Some are human-centered, and some are grounded in a connection with the land, sea, and other species. A big part of flourishing is translating happiness into your own terms. Luckily there's a lot of inspiration to draw from.

Taking cues from classical Western philosophy, climate happiness might take the form of *arete* (striving for excellence and meeting your goals), *eudaimonia* (a sense your life is meaningful and makes a positive difference), *epicureanism* (attaining tranquility, and the absence of pain and fear) or *stoicism* (accepting the world as it is). From the perspective of Buddhism, climate happiness can be a path from mental suffering and attachment to liberation and commitment to save all beings.

The field of psychology offers more ways to think about happiness and flourishing. These include *self-actualization* (or reaching your full potential), finding a sense of *flow* in a positive activity (whether working on a problem or spending time outdoors), the ability for *self-improvement* (the ability to take on new behaviors and habits), and the attainment of *wisdom* (balancing life experience and good judgment).[8] You can seek *liberation* by confronting mental health issues in the context of oppression and racism, or experience *post-traumatic growth* (finding new possibilities for life after recovering from a severe climate loss or injury).[9]

Adding Global Examples to Your Feelings Vocabulary

You can also find inspiration for climate coping in the words, phrases and concepts used to describe happiness and positive emotions in different cultures and languages. After all, if we all share the experience of the ecological crisis, we can also share our many visions of well-being. If you pick up positive psychology researcher Tim Lomas's *Translating Happiness* and just begin with the letter *A* you'll find words like the Sanskrit *advaita*, "non-dualism," the notion that the personal and the sacred are not separate. In Hawaiian and Polynesian cultures there is the warm greeting, and spirit of love and connection, embodied in *aloha*. In Italian *arrangiàrsi* means to get by with the best of your abilities. The Arabic *asabiyyah* signifies group solidarity and mutual aid.[10] You also find the Russian word *azart* to describe passion, risk-taking, and love of a challenge. All these variations on well-being can have a place in your coping and thriving.

In Native and Indigenous cultures, concepts for happiness and well-being often take some form of "walking in beauty" that includes healthy relationships, environmental sustainability, and connection to place. Some examples are the philosophy of *sumak kawsay* of the Quechua peoples of the Andes (*buen vivir* in Spanish), a collective sense of well-being that includes yourself, your social group, and nature. The Navajo ideal of *hozho* refers to existing in balance, peace, and harmony with the world. The Lakota worldview of *mitákuye oyás'iŋ* involves being in kinship and balance with an expanded sense of relationships. And we discussed in Chapter 12 the Māori concept of *turangawaewae* or "place to stand," where you feel rooted to the place of your ancestors and strive to protect it.

Make a point to search for words, concepts, or experiences that help you expand your possibilities for happiness and well-being. The more you look, the more you will find.

FINDING YOUR STYLE(S) OF HAPPINESS

Use the following set of questions to consider your approach to happiness and how this may influence your version of well-being and flourishing in a climate and environmental context. There are no right or wrong answers. But the list can expand your sense of possibilities and suggest a combination that works for you.

Do you consider happiness to be any of these?

- An inalienable human right to be protected
- A choice or form of human potential
- A measure of your horizon of hope and your ability to conjure the highest vision of the possible
- Your evolutionary birthright, as a human mammal, to experience a full suite of emotional experiences
- Something earned that comes from action, or as a reward for your service and effort
- Something you are granted by grace or a benign universe and that doesn't require any reason or transaction
- Random and a matter of luck
- Something useful and pragmatic that improves your health and well-being
- A positive personality trait that helps you be an effective leader or change-maker

Rather than viewing "climate change and happiness" as something to be wary of, consider how it might be a right, a grace, a freedom, a useful force. What permission can you give yourself to mobilize *your* version of happiness?

MAINTAINING HAPPINESS

Flourishing doesn't mean you have to feel great *all the time*. As we've discussed, coping with eco and climate issues brings up a range of emotions, high and low. On days when you are feeling particularly low, these strategies can provide the boost of positive emotions that you need to get you through.

Practice with Stretch Feelings

I asked Delphina to look at a list of feeling words and choose one that seemed like a far stretch. She laughed and pointed to "jubilation." I agreed that feeling jubilant (from the Latin meaning "a shouting for joy") might *seem* like a stretch given the troubling information she was learning. Yet I observed that I had seen eco-anxious people shouting for joy when they watched videos of their favorite animals, engaged in outdoor sports, or celebrated positive news. Delphina acknowledged that she could recall times like this. When she was a young girl, she screamed for joy when she first saw dolphins leaping in the ocean. She and her veterinarian friends cheered when the Illinois governor signed legislation prohibiting the burning of substances containing PFAS, and again when the US Environmental Protection Agency issued the first national, legally enforceable drinking water standards. If you identify and keep track of stretch feelings related to your eco and climate concerns, you may find they happen more than you think. Delphina realized she was a person who could still feel jubilant.

Be Aware of What You Take In

Earlier we talked about "staring into the abyss." One source of darkness is so common and obvious we don't see it. We are constantly taking in negative information. As we learned, only 3 percent of news about climate is good or solution-focused.

James lamented that the books he was seeing on climate change and the environmental crisis were depressing reads: *The End of Nature, Field Notes from a Catastrophe, The Great Derangement, The Uninhabitable Earth, Generation Dread, The Heat Will Kill You First*. If you pick up a book, start reading an article, or turn on a documentary and find that it is making you stressed, disillusioned, or exhausted, give yourself permission to just stop.

Read the Last Chapter

I also reminded James that the books he was selecting were meant to raise awareness and sound the alarm, not to comfort him. Thus negative information is stacked up at the beginning and solutions come later. After all, the authors were journalists and scientists, not therapists. I gave James some advice: If he was stuck, skip to the final chapter or the epilogue. This is often where authors share their vision for solutions, possibility, and flourishing— even if they might not use that term.

So as an exercise in hope and flourishing, take some time to pull the climate crisis books off your shelves, even ones you are familiar with or have had for years, and flip to the end.

BREATHE HAPPY

My first lesson on the value of breathing was in high school. Once during a sports physical I was told that my blood pressure was high and so I couldn't participate. I found a book about addressing blood pressure through meditation at the local library, and secretly practiced in my bedroom. At my next physical, my blood pressure was in the normal range and I was cleared to play. Later in graduate school, having never taken a yoga class, I signed up for an intensive course that met every day for a month. My instructor was a strict teacher, very focused on precise body posture and candid in his feedback ("Thomas, is that as deep as you can bend? Surely not!"). These classes taught me to pair conscious breathing and movement. Later, during my doctoral research, I came to see breathing, yoga, and body awareness integrated at the Mindfulness Meditation Program founded by Jon Kabat-Zinn at the University of Massachusetts Medical Center.

Larysa was doing much better after she had made structural changes in her life, like reducing her screen time and taking part in her local disaster-preparedness group. Now she was ready to work on reclaiming her nervous system. She had noticed that whenever she was immersed in disaster worries, her breathing turned into quick, panicked gulps (hyperventilation), causing her to feel light-headed and short of breath. From her birthing classes Larysa was already familiar with belly breathing, which involved taking deep breaths that engaged the muscles of her diaphragm. I coached her to reapply these skills to restore a sense of calm and well-being when she found herself overcome by climate anxiety.

To practice this yourself, place one hand on your stomach below your ribs and the other hand on your chest, and breathe from your body's center of gravity, from just below your navel. Practice inhaling and feeling the air expand upward through your abdomen and toward your collarbones. Then reverse the process with each exhale, letting your chest relax and slowly squeezing all the air out of your lungs. A metaphor Larysa used was that she was still "in labor" in some ways, this time birthing a new awareness of being a mother with realistic anxieties about the world. Looking at news stories, even troubling ones, just didn't hook her the same way when she was belly breathing.

When James was feeling sluggish, I encouraged him to use a yoga technique of "fire breathing." This technique paired relaxed inhales with active and forceful exhales using his abdominal muscles and had an invigorating effect.[11] To try this, purse your lips and use your abdominal muscles to exhale sharply from your nose. Imagine you are holding a candle and pretend to extinguish it by forcefully breathing air out of your nose. Fire breathing is a quick series of these forceful exhalations. James practiced this breath each morning, and any time throughout the day when he felt tired or lethargic. This helped him be more energized while maintaining a sense of calm.

Some years ago, Gurucharan Singh Khalsa, an American kundalini yoga expert, sought me out to discuss climate change therapy. When he discovered I was coping with my wife's cancer diagnosis, he offered to teach me some of his techniques for "breath-walking," a way of pairing yoga breathing and walks.[12] I shared some of those techniques with Christina to use when she was building her hiking practice. She began simply with five-minute sessions where she would take four steps for each inhale and four steps for each exhale, segmenting her inhale and exhale into four counts synchronized with her walking pace, mentally counting 1, 2, 3, 4 (like 4/4 time in music). Eventually she extended her breath walks and varied the pace of her breathing (a four-step inhale and slower eight-step exhale for relaxing and an eight-step inhale followed by a more rapid four-step exhale for energy and motivation). I suggested Christina pair fire breathing with her breath walks when she was approaching a hill or feeling tired. This helped Christina feel energized, adventurous, and confident—and more prepared for her pilgrimage on the Via di Francesco.

LIVE IN BALANCE: HOMO AEQUILĪBRIUM

The provocative twentieth-century environmentalist Edward Abbey is known for a quote from a lecture he gave to college students in the 1970s:

> One final paragraph of advice: do not burn yourselves out. Be as I am—a reluctant enthusiast...a part-time crusader, a half-hearted fanatic. Save the other half of yourselves and your lives for pleasure and adventure. It is not enough to fight for the land; it is even more important to enjoy it. While you can. While it's still here.[13]

This is excellent advice for when you are feeling a sense of guilt about the impact of your life choices on the environment. It reminds us to do what Sam the home builder we met needed to do: balance guilt with compassion, despair with joy, and commitment to one's values with the permission to experience happiness.

Nordic cultures have many concepts for balance, including the quality of *lagom*, a Swedish word meaning "sufficient," "enough," or "just the right amount." In relation to action, *lagom* can be translated as "in moderation," "in balance," "ideal," or "as much or as long as it takes." *Lagom* was really meaningful for Sam and his spouse, Liberty; it gave them a way to define flourishing for themselves and their family, and reminded them that a life of all work and no play was not a life in balance. In the age of climate change and the Anthropocene, we might begin to think of *Homo sapiens*, a "human being" that is "discerning, wise, sensible," as also Homo aequilibrium—the human that balances, and the human that is enough.

A MANIFESTO FOR FLOURISHING

Below is a statement that summarizes what flourishing in an era of climate crisis can look like. I want you to imagine this scenario for your life, and to imagine that it is realistic and attainable for you, right now. You can use it as an affirmation, or you could even read it as a personal manifesto. Take a deep breath and begin.

> I carry the weight of climate change on my shoulders—I feel anxious about the heat of the summer, the air quality, the winter storms, and the future. My internal alarm systems are going off all the time. But I can take a step back and acknowledge this. I'm conscious of how I'm thinking about the world and maintaining a growth mindset. I'm prepared to discover things are much worse than I thought, but also at times better. I'm practicing emotional intelligence. I stop and ask myself what I'm feeling— and also what feelings I want to have and to cultivate and grow stronger. I remind myself that sustainability begins with my own foundation of personal health. I'm strengthening my relaxation response so it can balance my stress response. I know that if I feel hurt about the state of the environment, it is a signal that I care.
>
> I tell myself, "Even as…" Even as I'm thankful for global awareness, I unplug and reconnect with my senses. I look for the news that is right outside my door, in my home, my community, and my local ecosystem. Even as the climate changes in unpredictable ways, I am developing a deeper awareness of my local weather, disaster threats, and my sources of risk and resilience. I keep learning. I find support.

I remind myself that climate change is not my fault. Even as I take on responsibility, I understand that forces of power, nationalism, and unregulated commerce are driving much of Earth's systemic troubles. Even as I feel like a hostage to dangerous forces, I focus on what I can control. I remind myself that political systems are created by people and they can be changed. When the media announces new threats, I remind myself, "I already know." I might not know every detail of the story, but I am familiar with the larger narrative.

I understand that many people are still coming on board with their own version of an ecological waking-up syndrome. I have compassion for them. Even as I validate their eco-anxiety, I'm not swept up in their troubled feelings.

Nature is not just a place outside of me. I am nature. I have a unique environmental identity, and set of experiences and values that give me a place to stand on the planet. Even as I am part of the web of life, I am also weaving it.

Even as I struggle with feelings of fear and eco-anxiety, I step back and see these as normal emotional responses. I know that standing up and publicly recognizing my eco-anxiety is a healthy form of self-expression. Regardless of who is listening, speaking my truth out loud generates energy for action. I am a work in progress, but I have new skills and am building a tool kit for coping. I know I am so much bigger than my climate anxiety. I can still feel happiness in this time.

Chapter 14

Relationships

Create around one at least a small circle where matters are arranged as one wants them to be.

—Anna Freud[1]

Coping, healing, and thriving in an era of climate anxiety is all about relationships. Nature is a system of relationships, of which we are all a part. Disasters are collective experiences that bind us in lifelong relationships with our fellow survivors, even the ones we never met. Your environmental identity, values, and eco-story all form and find expression within relationships. On a more abstract level, regulating your feelings is about regulating the *relationships* among your feelings, and how they conflict with or support each other. Having a healthy network of human relationships along with a healthy relationship to nature will make it easier to cope and recover during times when you feel anxiety and despair.

Thriving in an age of climate anxiety is all about relationships too: maintaining them, growing them, and standing up for them. Standing for your values and your home place supports your ethical flourishing, both your pleasure in life and your *eudaimonia*—living a life that is virtuous and fulfilling. Finally, the communication skills, connections, and group alliances that grow out of your relationships will set you up for more effective action.

However, relationships do not come without challenges, especially in today's polarized era, when people close to us can hold divergent views about the causes and impacts of climate breakdown and the urgency with which we must address it. This chapter addresses relationship skills for a world on fire. It will arm you with the tools to form connections across opposing values and viewpoints and be more confident and secure in relationships with individuals coming from different parts of the climate elephant.

FAMILY DYNAMICS, POLITICS, AND CLIMATE

When Delphina began studying the effects of PFAS, the "forever chemicals," in her Great Lakes region, she badly wanted to tell her family and neighbors about what she was learning. But whenever she brought up the issue of industrial chemicals or climate change, and the associated social and political differences, people found a way to redirect the conversation. This is known as *social silencing,* a form of denialism.[2]

When Delphina pushed the issue, people responded in a way that actively shut down the conversation. To help her understand some of these dynamics, I shared some roles people take on during polarized political discussions, identified by family psychologist William Doherty.[3]

- *The sniper.* Someone with a passive-aggressive style and the habit of making snide comments or inflammatory points tinged with humor or teasing.
- *The gladiator.* The person who is passionate and always looking to debate.
- *The defender.* Someone who springs into action if their typical ways of doing things are questioned.
- *The diplomat.* The one who tries to smooth over differences.
- *The bystander.* Someone who remains silent.

In Delphina's family, she was the gladiator, eager to talk about her research and the urgency she felt. Her politically conservative uncle was the sniper. When Delphina became upset by something he said, he deflected the focus back to her with the rejoinder, "Can't you take a joke?" Her father was the diplomat, eager to quell the argument. And the rest of her family members were bystanders, unless someone felt their values or beliefs being threatened. Then they would get drawn into the role of defender, arguing for their side. Welcome to Sunday dinner.

Sometimes entire families take on a dominant style. For example, Roman's soft-spoken, conflict-avoidant family was full of diplomats, whereas Christina's demonstrative, opinionated family was full of outspoken gladiators. Think about how these patterns may play out in your family or social groups you belong to. What role do you take on?

Delphina assumed that her family simply didn't care about "forever chemicals" or the climate. I encouraged her to imagine some other possibilities. As psychoanalytic theorist Renée Lertzman suggested in her groundbreaking essay, "The Myth of Apathy," an apparent lack of caring about the climate crisis can often be a psychological defense mechanism against overwhelming emotions. We "unconsciously deny what is staring us in the face because what is at stake is too painful to consider," she noted.[4] Yet on the surface, this can look a lot like apathy, disengagement, or what Lertzman called "environmental melancholia." Other times, as Joanna Macy and Molly Young Brown observed, people avoid bringing up environmental or social issues out of fear, including fear of seeming uninformed or unintelligent, fear of appearing morbid, fear of guilt and personal accountability, fear of causing distress or burdening others, and fear of appearing weak and emotional.[5]

Lertzman's insights gave Delphina new ways to understand her family and community. In addition to recognizing the influences of differing values about nature, she was better able to appreciate that her family sought to avoid conflict as a form of emotional self-preservation: a way to show love, maintain contact, and keep the peace.

When I meet someone who is detached or resistant to addressing climate change, I know that it doesn't necessarily mean they don't care; it simply tells me that they care *about something else even more*. This assumption turns everything around. I stop thinking of the person as uncaring and become curious about what it is they do care about. And that is a much more constructive way to start conversations. This insight led Delphina to step back and consider all the other reasons she and her family struggled to find common ground on climate and ecological threats: different levels of education and environmental knowledge, hearing disputing messages in the news, and competing priorities.

As the old saying goes: "People don't care what you know until they know that you care." Delphina found that when she spent time searching for what her family members cared about—her uncle, her father, her working-class cousins—it helped her see the issue from their perspective, which in turn made them more open to hearing about hers. Then she could talk about her research on PFAS and start to share some of the information she knew.

COUPLES AND CLIMATE CHANGE

Bad feelings, unhealthy behaviors, or relationship conflicts can arise in relation to your fears and concerns about eco and climate issues. These can range from small disagreements about daily acts to deal-breaker choices like whether to have children. As Panu often notes when we discuss these conflicts on our podcasts, these issues can be particularly fraught because they are not simply "lifestyle choices" but rather "life-constituting choices."

Like other emotional or behavioral problems, however, relationship conflicts rarely occur randomly. They have patterns, like happening at certain times of the day or in certain places. The key is to find these "high-risk zones" and bring extra energy and attention to them. Sometimes small changes can make a big difference.

To find your own high-risk zones, look across your weekly calendar and place a red circle around hours or days when you tend to have conflicts with your significant other, children, or even housemates. Issues may cluster around the morning routine, as they did for Liberty and Sam; the transition out of "work mode" at the end of the day, which was difficult for Marcus; or later in the evening, when tired couples are trying to relax and reconnect, as was the case for Roman and Sandra.

When we're debating with our significant other about some ecological behavior or political stance, we're really arguing about how we're showing love to ourselves and the planet. A conversation prompt I find useful during these moments of high tension is the following: "My love for nature is conflicting with your love for nature in some way. And then it starts to conflict with my love for you and your love for me."

HOW TO LISTEN

Stephen Covey famously wrote in *The 7 Habits of Highly Effective People*: "The biggest communication problem is we do not listen to understand. We listen to reply."[6] To take Covey's insight one step further, in a polarized world of climate breakdown, "listen to reply" easily becomes "listen to attack." Too often, we find ourselves formulating a rebuttal even before the other person is done making their point.

Here are some basic ground rules for conducting healthy, productive conversations with people who don't necessarily share your view on ecological issues.

- Avoid a "harsh start-up" to conversations. Catching people off guard will invite defensiveness and pushback. Communicate the need to talk about important topics in advance, and pick a time.
- Be attentive to the *levels of communication* you are using. A common problem occurs when one person shares their feelings or concerns and the other responds with facts or opinions. Are you making small talk? Discussing basic, agreed-upon facts and information? Debating differing opinions and beliefs? If another person is talking about their feelings and you are making small talk, the other person may feel unheard or invalidated.
- Practice "co-regulating" emotions by validating the other person's feelings and making clear that you are not giving up on the relationship, even if you are irritated and need a time-out.

For example, Martin found it helpful to look Sally in the eye and hold his hand while they worked through their respective differences on issues like recycling.

- Try to extend a *growth mindset* toward your relationship. It's easy to form an opinion about others and seek evidence to confirm it. A paradox in relationships is that our inherent need to grow and expand ourselves takes place in close relationships where we often know the other person as well as we know ourselves. So we need to be open to revising our view of the other person to give each other room to grow.

- Work to be flexible and accept the inevitable stylistic differences that exist in all relationships. One person may be quiet, while the other is expressive, or one person might have grand visions, while the other is focused on the practical details. Recognize that each style has something to offer.

- Be aware of how you and the people you care about give and receive love. Some prefer comforting words. Others look for concrete actions. Others seek alignment on values and beliefs. Keep the other person's "love language" in mind when dealing with debates about environmental concerns.

- Be mindful that everyone has their own priorities, even among groups or families who share common values. While two people might both value biodiversity, for example, one may prioritize natural beauty (aesthetic value) and the other may prioritize science (intellectual value).

"CLIMATE CAFES" AND THE BENEFITS OF GROUP SUPPORT

As a therapist, Rachel was empowered by the insights she gained into her own eco-anxiety about her daughter's move across the country for college. Moreover, the positive experience of sharing her concerns openly revolutionized her approach to practicing therapy and inspired her to join other climate-conscious therapists in leading public "climate cafes"—open discussion groups that invited people to share their questions and concerns about climate breakdown and other issues.

As I helped Rachel prepare for her first climate cafe, we reflected on the healing properties of being in a healthy and supportive group. For one, it's easier to bear sadness about lost seasons or species when you also feel the compassion and acceptance of the people around you. We humans did not evolve as solitary beings. Our impulse to *tend and befriend* in response to stress is an interaction of oxytocin and dopamine just as real and hardwired as the fight-or-flight response.[7]

Like many mental health professionals, Rachel and I were both influenced by psychiatrist Irvin Yalom and his classic research on the benefits of group therapy (and many other kinds of support groups). As you read the list below, notice how these positive relationship processes take on a new significance when coping with eco and climate distress.

1. *Universality.* Learning you are not the only one with your type of concerns.

2. *Installation of hope.* Seeing tools or concepts that work for others with similar problems, giving you reason to hope that they could work for you too.

3. *A sense of belonging.* Being accepted by the group and developing bonds and trust.

4. *Emotional relief after sharing strong feelings.* Being able to say what is bothering you, and feeling better once you do.

5. *Opportunity to practice communication skills.* Everything from basics like sharing, listening, and taking turns to more advanced skills like giving and receiving feedback, bestowing compliments, and airing differences or grievances.

6. *Vulnerability.* Gaining confidence to disclose your private thoughts or feelings.

7. *Spiritual and existential awareness.* Facing up to the realities of life and death, clarifying your most important priorities, and being less caught up in trivialities.

A SMALL CIRCLE

Christina was fortunate enough to have several circles of people in her life who shared her climate values. The core band of political organizers she had worked with for many years was one. Her family was another: Her extended family lived in the same Philadelphia neighborhood as Christina and in several cases on the same street. For her, flourishing meant nurturing and strengthening the small circle of relationships that helped her feel less alone in her climate worries.

ECO-FRIENDS

One source of strength and inspiration I have relied on is my network of what I call my *eco-friends*, all the people I know who demonstrate environmental values and do good things for humans and Earth. Though I haven't stayed in close touch with all the people I met working for Greenpeace in the early 1990s, in wilderness therapy, or river guiding in the Grand Canyon, I know they are out there living their environmental identities. One is a builder of sustainable homes. Another is a filmmaker who does projects for their state's department of conservation. One friend restored an old bar in a logging town that became a new community gathering place. Some run nonprofits. Others are artists. A number are researchers, scientists, and teachers. And this doesn't even count all the people studying or practicing psychology, climate therapy, and sustainability whom I have worked with over the years. Add the listeners to my podcast who tune in from thirty-five countries, and this network ultimately numbers thousands of people around the world. And now that global community also includes you and all the other readers of this book.

Earlier I referenced environmentalist and entrepreneur Paul Hawken and his 2007 book, *Blessed Unrest*, which drew attention to the thousands of different groups around the world dedicated to environmental sustainability and justice. Recognition of this diverse but universal human impulse to improve the state of the planet was an important paradigm shift for me.

After reading Hawken's book, I never truly felt alone in my work. Now, when I consider my varied eco-friends, I feel like I have my own version of a global community.

Perhaps you have a diverse network of eco-friends too. If not, I can predict that when you begin to take action, you will eventually build one.

CELEBRATING DIFFERENT WAYS OF THINKING

For Braden, the forester from Oregon, taking action on issues of climate often required joining forces with a surprisingly diverse crowd of friends and colleagues, whom he summed up as "cowboys, hippies, rebels, and Yanks, along with Natives, science nerds, yuppies from the city, wise users (focused on drilling, mining, and harvesting), and Earth-firsters (who wanted the land left alone)." In his work, he was just as likely to speak to a researcher from the Yale School of Forestry as to a foreman at a local sawmill about preserving small-town jobs.[8]

Braden had learned the value of seeking diverse perspectives from his long-time collaborator Ennis, whom he had known since their days at the forestry school at Oregon State University.[9] Braden and Ennis had worked together for years, but they were also very different. Braden was fascinated by wood science and believed in sustainable forestry—the idea of keeping some mills operating and people in jobs while still preserving the natural integrity of forests. Ennis was an unapologetic tree-hugger. He put ecology and biodiversity above all and believed in harvesting resources only after the wilderness had been preserved for people to experience. In forestry school, the two friends had adopted the personas of famous environmentalists. Ennis was the "John Muir" and Braden was the "Gifford Pinchot"; Pinchot's pragmatic philosophy was often pitted against the preservationist views held by Muir, the nation's most prominent naturalist. Another friend was their "Edward Abbey," fiercely speaking out against fossil fuel pipelines and coal-powered trains and advocating "no compromise in defense of Mother Earth"–style activism. Another was their "Gary Snyder," a Buddhist and poet who led wilderness retreats. Collectively, they were hunters, vegans,

Luddites, and techies, yet all were united by their love of the woods and their stubborn respect for one another. They were hard on each other, but also fiercely protective of each other's views and the right to express them.

ECO-FRENEMIES

The *frenemy* (a person whom you are friendly with but who is in some ways an enemy or rival to you) is a relationship to be aware of when you are doing climate or environmental work. Many times people who are on the same team in the broad sense of supporting protection of the environment and taking action on climate breakdown can find it hard to collaborate because of deeply held beliefs (or dogmas) about what constitutes correct action or strategy. As we learned in Chapter 6, climate action can be motivated by a wide range of diverse and sometimes opposing values. Add to this many different professional backgrounds and worldviews; race, culture, and class differences; and individual personality and stylistic differences, and it's inevitable that frictions would arise.

When it comes to frenemies, it helps to remember that while you are not always on the same team, you're at least in the same league. You can celebrate differences and coexist, find mutually beneficial coalitions, or give people space to work on their own projects, trusting that you are all heading in the same direction. Part of the process of collaborating is learning to work around disagreements and rivalries.

COMMUNICATING ACROSS THE AISLE: BRAVER ANGELS

In highly polarized situations, disagreements become more pressurized. Amid stark political divides, one or both sides can negate each other's position, or even their right to be a part of the discussion. Facing a gridlocked state legislature, Christina realized she would need to learn how to understand and communicate with political conservatives (including some extreme right-wing elected officials) so that functional climate legislation could be passed. She lamented that she had no close friends or confidants with experience in right-wing politics from whom she could gain an understanding of a conservative perspective. So I directed her to Frank Luntz.[10]

As a political pollster, Luntz became famous (some would say infamous) for helping the US Republican Party formulate talking points downplaying the urgency of climate change. In recent years, however, his work has shifted dramatically. He has taken up the opposite challenge of helping Republicans and Democrats craft effective messaging about the need for climate action. Luntz's approach involves personalizing, individualizing, and humanizing the impacts of climate change to make these issues more relatable to the average citizen, while also focusing on the benefits and opportunities of climate actions. Luntz notes that the war over the reality of climate change in the United States is effectively over, with a majority of voters on both sides of the aisle acknowledging the issue.

His insights helped Christina *get out of her own mindset*. She learned to replace terms like "sustainability" with "a global commitment for a cleaner, safer, healthier future." And rather than leading with scientific data about consequences, she learned to start positively with the personal benefits of action, state consequences, and end positively: a "solution sandwich." She described the policies she was advocating as "fact-based" rather than using the more academic term "evidence-based." As Luntz noted, we can argue on evidence, but we can agree on facts. This messaging would more effectively address the questions of conservative voters: "What is in it for me or my family? How does America benefit? What are the consequences of inaction or wrong action? And why should I trust you?"

Empowered by her new perspective, Christina made a commitment to get to know more people on the other side of the aisle. She turned to Braver Angels, a cross-partisan, volunteer-led movement in the United States that works to overcome *emotional polarization*—the process of society separating into antagonistic groups (like "progressives" and "conservatives") that do not trust each other and cannot work together. The method uses group discussions and social events designed to build community and create action at the grassroots level.[11] As Christina found, discussions between political groups worked best when they had very clear protocols and sequences. Each took turns discussing their values and goals, as well the issues and problems they saw on *their* own side, while the other group quietly observed. Then the two groups came together to discuss commonalities.

These dialogues helped Christina create a road map she could use to listen and respond to anyone across the political divide.

AGREEING TO DISAGREE

Earlier, when we discussed the importance of people's eco-values, I evoked the poetry of W. B. Yeats: "Tread softly because you tread on my dreams."[12] Naturally, people want you to join them and are frustrated if you disagree. It can seem impossible to find a place beyond *their side* or *your side*. But there are ways to be present and accepting of people without agreeing with them or taking up their cause (or negating them altogether). Here are two strategies for navigating those difficult, high-pressure conversations:

1. *Taste the bait, but don't swallow the hook.* When you begin to work across the partisan divide, people are often trying to either recruit you to their side or hook you into a debate. I like to say, "Taste the bait, but don't swallow the hook." When you are confronted with a different belief or opinion, be secure enough in yourself to hear the person out. "Taste the bait" by listening and trying to discern the needs or meaning underlying the person's political position. But don't feel obliged to agree or let it hook you emotionally. You don't have to agree on everything to work together with someone. But you can offer respect, curiosity, and genuine interest.

2. *Show your work.* Rather than pushing people to change their conclusions, it can be helpful to show others how you came to yours. That's where the "I felt, I found, I feel" technique comes in. The basic template is a variation of "Originally, I felt _____ about [topic] _____. What I found was _____. So now, I feel _____ about _____."

Earlier, I talked about the notion of a "personal carbon footprint." When I first became engaged with global warming, I naturally gravitated to solutions dependent on individual action because that's what I felt comfortable with as a psychologist: personal change. What I found was that climate change was really a structural, economic, and cultural problem. While individual actions are important for many reasons, our options are quite limited given the systemic rules we are forced to play by. Moreover, focusing on personal responsibility was effectively colluding with climate change propaganda and exacerbating the climate hostage situation. This went against my ethics as a psychologist. Now I feel that my understanding of the problem and the approach we are taking in this book is more honest and accurate.

By admitting that you didn't always have all the answers and being open about the lessons you've learned along the way, you give permission for others to learn and grow too.

RECOGNIZING UNHEALTHY GROUP DYNAMICS

Although collaborating with people who hold diverse viewpoints on eco and climate issues can give you insight into a broad range of values and perspectives, it would be a mistake to assume that all group processes are healthy and safe. In fact, some group dynamics can make disagreements even worse. As Larysa gained more experience with disaster preparation, she recognized unhealthy processes in some of the online disaster-prepping groups she had frequented. Negative group processes can include:

- A preponderance of negative emotions, leading to a downward spiral of mood. Emotions are contagious.
- A pattern of recycling the same information, and getting stuck in collective blind spots. Failing to question assumptions or invite new perspectives creates a self-perpetuating echo chamber.
- Members who spread misinformation (inaccurate facts and ideas) or, worse, pass on disinformation (purposely misleading or harmful information).

When Larysa contrasted these dynamics with those she observed while working with her local Community Emergency Response Team (CERT), she could clearly see which group styles were primarily healthy and positive. High-performing groups like Larysa's CERT team tend to share the following characteristics:

- They have a variety of members who don't always agree on everything.
- People share new and unique information (instead of recycling the same ideas).
- People feel safe questioning assumptions and asking hard questions (instead of engaging in social silencing or groupthink).

High-performing groups don't happen by magic. They result from clear intentions and norms, the efforts of members, and having leaders that support these processes and role model them themselves.

BEING OTHERISH

Ecological threats can feel so big and urgent that sometimes we respond instinctively with selflessness, throwing ourselves blindly into action. Although this might sound like a virtuous impulse, unchecked selflessness can quickly lead to burnout. In his studies of altruism and volunteering, organizational psychologist Adam Grant notes, "I've consistently found that self-interest and other-interest are completely independent motivations; you can have both of them at the same time…If takers are selfish and failed givers are selfless, successful givers are *otherish*: they care about benefiting others, but they also have ambitious goals for advancing their own interests."[13]

Christina found that the idea of being *otherish* allowed her to make space for her hiking and Via di Francesco pilgrimage while still having enough energy to devote to her political advocacy on behalf of eco and climate issues. The concept of "otherish" is another example of how we can hold competing emotions or motivations and find a healthy middle ground (this is the dialectical thinking we discussed earlier). Just as you can understand that the climate crisis is both worse and better than you think, and just as you can feel both sadness and inspiration about ecological issues, you can be motivated by both self-interest and concern for others.

IT'S NOT ALL ABOUT YOU

I have a cartoon by the illustrator Sarah Kempa pinned to my office wall. It's a movie poster for a hypothetical film called *Not Actually About You at All*, with a tagline that reads, "Critics say: 'Stop projecting yourself onto everything.'" This gentle humor reminds me not to take myself too seriously, and to remember that not everyone shares my same needs and concerns about the state of the world. By definition, relationships are *not all about you*. Relationships are about seeing others, including the ones without a voice, future generations, and even (or perhaps especially) the people who disagree with you.

If we widen our relationship lens out further, beyond the human sphere altogether, it's actually not about us *humans* at all. It's about the more-than-human world and the larger body of Earth. Sometimes we need to stop and reset our awareness and consciousness in order to pierce the human-centered bubble most of us live in.

"Not actually about you at all" also reminds me that at times, good actions—like behaving well toward others or caring for and defending Earth—will go unrewarded, unrecognized, and unappreciated. You might be criticized, blamed, or even threatened, punished, or hurt. When this happens, you may believe you are not seen. But *I* see you. I may not know you, but I know you are out there. You are not alone.

Chapter 15

Art

What thou lovest well is thy true heritage.

—Ezra Pound[1]

It was a rainy day in Portland when Slade came to me feeling burned out, anxious and distracted. Slade was a sustainability director at the local port authority, which oversees a large swath of infrastructure integral to the area's economy, most of which is either at risk from climate upheavals or requiring updates to ensure a safe, low-carbon future. Beyond the ocean and river shipping terminals, Slade's sustainability mission extended to several airports, rail hubs, business parks, and shopping areas, as well as many acres of critical bird and fish habitat. As an avid boater and fisherman, he loved being on the water. Slade saw himself as a riverkeeper. He was proud of the fact that finally, after a century of neglect and dumping, the waterway flowing through town was clean enough to swim in.

Slade had come straight from a grueling local government meeting, where he had learned that 2,500 miles of sewer pipes in the city, many a century old, required upgrades to protect drinking water and salmon habitats from the storms and flooding of a changed climate. Slade had always considered himself more a "doer" than a "feeler." He had originally sought help to manage his adult ADHD symptoms, but lately he felt overwhelmed by the emotional weight he carried. At times, the mission of his job seemed impossible, balancing constant pressure for economic growth and expansion while striving toward sustainability and helping shepherd the region through the climate crisis.

Now Slade was sitting in my office with his eyes closed, trying to focus on sounds streaming through headphones. The music I had chosen for him was a recording of "Become River," a fourteen-minute composition by contemporary composer John Luther Adams.[2] The Seattle Symphony begins, almost imperceptibly, with tinkling chimes evoking the vibrations of ice crystals high in the atmosphere or perhaps interred in alpine glaciers. As the piece progresses, the orchestration of the strings and horns evokes dripping rivulets of spring meltwater and the trickle of mountain streams. The performance builds in intensity toward a crescendo that brings to mind waterfalls and immense rivers flowing into the sea, then closes with darker tones, intimations of rising seas and flooded valleys. When the music stopped, Slade removed the headphones and opened his eyes. He took a deep breath. Then he had a moment of clarity. The relationship between himself and nature was as fluid and interconnected as the rivers he was dedicated to. He could see that his sense of urgency around ecological issues, compounded by his ADHD tendencies, would produce the jarring mood swings he was suffering.

I had Slade revisit his feelings vocabulary list and check in on the stretch goal he called "slow feeling." Listening to the music, he could sense his emotions rising and falling, ebbing and flowing, much like the tide he loved to observe in the local rivers. He was patient and focused. The work was waiting.

In this chapter, I'll show you how music, art, and creativity can be a prescription for coping, healing, and thriving in an era of climate anxiety. Given the gravity of the situation, it's easy to think that focusing on the arts is frivolous. However, it is through art—recognizing, appreciating, and creating it—that you can express your feelings and identity, recover your ability to be inspired by the beauty in the world, and create a state of flourishing in your life.

I encourage you to approach the arts in a way that is broad and open, that includes everything from the fine art you might discover in a museum or symphony hall to a culinary masterpiece you create at home or a video game with graphics you find awe-inspiring. The way you express your creativity can be anything from a formal artistic gesture that you've practiced to casual doodling, singing in the shower, or strumming a guitar.

Personal sustainability includes the art of living well, of finding beauty— in your neighborhood, your home, what you eat, how you dress—and finding moments of contentment and *joie de vivre* (simple delight in living your life). As we'll see throughout this chapter, the arts serve as a catalyst for powerful emotions, as life-changing experiences, and as a pleasant distraction from troubling realities.

ART HELPS FORM YOUR ENVIRONMENTAL IDENTITY

Starting from childhood, art can influence your relationship with nature in profound ways you may not even be aware of. Look back on your life and the eco-timeline exercise from Chapter 5. You'll likely find children's books, novels, pop songs, movies, TV shows, poems, paintings, or videos where you encountered images of nature and environmental role models that formed the building blocks of your environmental identity. We saw how the painting *Christina's World* continued to give Christina new ways to see her own world throughout her life.

When Delphina was a young girl and visited her grandfather, he would take out his big, illustrated book about the history of the ancient Greek world and show her pictures of the ancient dolphin frescoes in the Palace of Knossos on the island of Crete.[3] "This is your name!" he would always say to her. "Delphina is the dolphin!" He gave a girl living in the suburbs permission to love the ocean, and to see herself in other animals. When she was older Delphina was excited to see the dolphin frescoes of Knossos appear with her archaeologist heroine Lara Croft in her favorite video game, Tomb Raider.[4]

Music was a building block of Rachel's environmental identity growing up in 1970s Boulder, Colorado. She learned to play the acoustic guitar from her mother and father. Her parents were a fiercely independent, biracial couple who partnered briefly but did not marry. Strong environmentalist values and love of playing music were things the young family shared. Folk singer Kate Wolf's song "Brother Warrior," which told of Earth waking to the dawn of another age, was a favorite of her mother's.[5] Her father was known for his rendition of Marvin Gaye's anthem "Mercy, Mercy Me (The Ecology)," one of the first Top 40 songs to openly address environmental problems. To this day, songs like these remind Rachel of the precious moments of warmth and closeness she felt with her parents, despite her distance from them now.

Marcus remembered how nature shows like *Wild Kingdom* expanded his childhood and how he wanted his own children, whose daily experiences were mainly urban, to learn and love the wild.[6] It was hard to get them to sit still for his favorite BBC documentaries with their gorgeous cinematography, but the whole family could enjoy watching Baratunde Thurston's show *America Outdoors.*[7] It was important for Marcus that his kids see a Black man enjoying outdoor adventures in all parts of the United States while being a leader who cared about the natural environment.

Exercise: Reflecting on Your Artistic Influences

What artworks or creative activities are integral to your sense of what nature is and how you think of your environmental identity? When did you first become aware of a species with which you closely identified through music, TV, or a movie? Was there a photograph or a painting that pulled you in and made you feel at home in nature? How does art influence the way you understand issues like climate breakdown, disasters, or other environmental problems, and ways these can be solved? Take a moment to think about all the ways arts have influenced your sense of who you are in relation to nature and Earth.

ART HELPS YOU FEEL BETTER

In addition to helping form your environmental identity, arts and creative acts can help you "feel" better about eco and climate issues, whether by bringing you inspiration and awe or as a way to express climate anxiety or grief. Oregon artist and poet Daniela Molnar does this by using the natural pigments she gathers in her wilderness journeys in her paintings. Daniela has described how she simultaneously confronts forces of grief and wonder in her creations and sees art as alchemy that can transform our emotions about environmental problems. Using art to express painful feelings about Earth is "simultaneously an injury and a process of healing."[8]

When feeling stressed, creating art is a quick and effective form of emotional regulation, helping you stay calm when you take in new information. Coloring, sketching, or knitting provides a reliable increase of blood flow to the *medial prefrontal cortex*, your brain's reward center. The poignant music you listen to and lyrical poems you read provide an opportunity for catharsis—a feeling of emotional release, especially of negative emotions such as grief and anger. Art also comforts us. Looking at pictures of cute animals and beautiful nature scenes can take your mind off negative emotions and help you redirect toward positive ones.

The arts can help you feel better not just by lifting your mood but by literally strengthening *your ability to feel* and expanding your capacity to hold powerful emotions that come with climate breakdown, such as terror, feelings of betrayal, and outrage.[9] This makes it easier for you to see the world as it is and bear witness—to tell yourself "don't look away" when you have the impulse to avoid or deny.

Art therapy, which encompasses drawing, painting, sculpture, music, and dance, has been a formal discipline for nearly a century. There isn't (yet) an established field of climate anxiety art therapy, but you can create this for yourself. The music therapy I was doing with Slade, for example, used sounds and rhythms to influence consciousness and healing. Music affects similar brain circuits and neurotransmitters as the anesthetics used in surgery. Neuroscientist and music producer Daniel Levitin has documented the benefits of music therapy for issues ranging from stuttering and ADHD to trauma, PTSD, depression, panic disorders; it also can play a role in managing physical and emotional pain.[10]

We reach out to art like a lifeline. Poems like Mary Oliver's "Wild Geese," with its advice that "you only have to let the soft animal of your body love what it loves," have long been a solace for nature lovers.[11] In the 1995 film *Il Postino*, Mario Ruoppolo, the eponymous small-town postman, tells the visiting poet Pablo Neruda that "poetry doesn't belong to those who write it; it belongs to those who need it!"[12] There are whole literary movements revolving around the grief and loss associated with destruction of the natural world.[13] As poet Kim Stafford said, echoing British writer Robert McFarlane, "A landscape that has not been evocatively described becomes easy to destroy."[14]

ART HELPS YOU SEE

Art and creative activities are vitally important because they help us see the reality of issues like global climate breakdown, extinction, and plastics pollution in ways we can't otherwise, allowing you to imagine possibilities, including worst-case disaster scenarios, and test yourself against them. Just like the concept of the "lost possible self" regarding ecological grief, appreciating and doing art can deepen your wisdom and growth mindset by exposing you to different perspectives, allowing you to better grasp the nuances of your eco-emotions and values.[15] If handled properly, art can be like the exposure therapy Larysa used to gain control over her climate anxiety triggers.

Reading stories and novels about environmental issues, like climate fiction, or "cli-fi," helps us empathize with others around the world who are already struggling. You can read about a fictional heat wave that kills millions in India in Kim Stanley Robinson's *The Ministry for the Future*, or a hurricane storm surge that traps young Natalie Torres in her house in Alan Gratz's young adult novel *Two Degrees*, and imagine how you might cope.[16]

Arts also help you see what people very close to you experience and value too. The only way I could get a felt sense of what Roman experienced in his time on the polar ice sheets was to witness the collapse of a massive city-sized glacier in Greenland through the work of photographer James Balog.[17] Martin was finally able to take in the scope of his husband Sally's concerns about plastic pollution when he saw images of the guts of dead seabirds filled with plastic waste that were captured by conservation photographers. Jesse, a Euro-American woman who grew up in the forests of the Pacific Northwest, saw her own environmental identity development reflected in Erica Berry's memoir *Wolfish*, and felt seen in Berry's words about the ways climate fears are handled in intimate relationships and how easy it is, especially for women, to take on polite socialized denial and stuff down their worries.[18]

Art also helps us appreciate the *multiverse* of sentient beings we share Earth with by showing us how *they* see the worlds they inhabit. When I was a boy growing up in Buffalo, more than 1,500 miles from the nearest coral reef, experiencing *The Undersea World of Jacques Cousteau* on television was the only way I could identify with the wondrous coral gardens, fishes, and octopuses along the Great Barrier Reef—and have reasons to protect them and their home. New generations of children can now view nature videos from all over the world.

When accessed safely and with care, creative arts and entertainments are literally visionary for us, bringing the abstract hyperobject of climate change and the momentous changes of the Anthropocene into real perspective. Arts can help you overcome your scale vertigo and develop the ability to locate where you can take action on a problem that is local and touchable in space and time. In his classroom, Reid was able to help his students understand what a phrase like "the end of nature" means in real time by having them access architect and sculptor Maya Lin's interactive online project *What Is Missing?*, her memorial to the sixth mass extinction of Earth's species.[19] Visitors can click anywhere on the site and learn the history of a place, what species have been lost, and those currently at risk.

Moreover, art can help you see solutions, which is why appreciating, supporting, and creating art goes hand in hand with science and activism. As neuroscientist Daniel Levitin writes, "We may think of science and art as standing in opposition to one another, but they are bound by a common objective. Science seeks to find truth in the natural world; art seeks to find truth in the emotional world."[20]

ART HELPS YOU RE-SEE

Art can help you *re-see* and relearn climate and nature history, while helping you shed light on your eco-timeline. I recently sat in on a discussion between my local mountain climbing club and musicians from the Portland Symphony who were restaging composer Richard Strauss's 1915 *An Alpine Symphony* as part of their series The Nature of Music.[21] Conductor David Danzmayr explained how Strauss's work evokes a day's ascent to a high peak in the Alps: daybreak and exultant sunrise, a slow but steady climb among alpine meadows, a passage through treacherous glaciers, and a transcendent ascent to the summit, followed by a rocky and tired descent until the final notes of the denouement at sundown. One hundred years later, the symphony remains a wonderful metaphor for life. But it also took on new meanings for the climbers and musicians. With the disappearance of glaciers in the Alps and the nearby Cascade Range looming large in their minds, they can no longer hear the same symphony. In the shadow of climate breakdown we interpret the Alpine Symphony—*and maybe all art about nature*—differently. Works that include a "calm before the storm" seem more ominous; elegies and laments are more poignant and closer to home.

ART HELPS YOU TAKE ACTION

The power of art can give you the courage to stand for your own voice, vision, and environmental values. Delphina developed her interest in animal rights and veganism, and learned of biophilia, humans' evolved tendency to love and connect with nature, not through her science classes but through the works of musical artists she admired like Moby and Björk.[22]

Appreciating, supporting, and *doing* creative work is an active form of coping with climate anxiety and despair. Not only does it foster a growth mindset and expand your possibilities for action, it can also serve as an effective environmental action in itself. For example, in her collection of writings on positive climate futures, *What If We Get It Right? Visions of Climate Futures*, Brooklyn, New York–based marine biologist Ayana Elizabeth Johnson places Blackness and Afrofuturist art on equal footing with climate science and activism.[23] The artist Olalekan Jeyifous brings climate adaptation to life in photomontages of an alternative, tree-lined Brooklyn that is vibrantly sustainable and multicultural, and linked by rail with "proto-farm communities of upstate New York" to reconnect city dwellers with the land and local food production.[24] In a time of global climate disruption, being an artist—maker, creator, or builder—activates the healing properties of building community, and provides an opportunity to be a teacher and leader in society.

Some artist-activists create large-scale works that involve entire cities and communities. Connecticut artist Mary Mattingly planted a vegetable garden and fruit trees on an old hundred-foot-long river barge and towed it to neighborhoods in New York City, where residents could gather as much fresh produce as they wanted for free.[25] In addition to being a thing of beauty, Mattingly's floating installation brought attention to the issue of urban "food deserts" (places that lack access to healthy fresh food) and antiquated New York City laws that prevent people from foraging wild herbs and fruit from urban trees.

ART HELPS YOU LOVE

As author Erica Berry noted, we need stories about climate futures to "help us visualize how to better love during crisis…[I]t's the love stories that make our bodies feel like we are there and…motivate us to fight for a better world back home." In truth, love stories are just what we need to envision positive futures. We need to know that people survive, that relationships survive, that love survives.[26]

MAKING YOUR OWN ART

Engaging in the creative process is beneficial to your physical and mental health. Painting, writing, knitting, woodworking, baking bread, singing in the shower, choreographing a dance, or making a collage with found objects can all be bricks in your sustainability pyramid.

You don't need any special skills or talents to make art that helps you cope with climate anxiety and other big feelings. Don't judge yourself. Just create. As children's book author Joan Walsh Anglund said, "A bird does not sing because he has an answer. He sings because he has a song."[27] As Braden, the forester drawn to woodworking found, making something beautiful or useful with your hands is healing in itself.

THE ART OF THE CLIMATE PLAYLIST

Readers of a certain age will be familiar with one of the biggest technical innovations in popular music, the mixtape—or its more modern iteration, the curated playlist. My podcast partner, Panu, and I have a ritual of sharing "climate playlists" with each other as a fun way to keep in touch. To express cynicism about an unjust system and being a climate hostage, Panu suggests the righteous elder doom of Black Sabbath's "Age of Reason."[28] I segue to the cool syncopation of the Weather Station's introspective critique of global capitalism, "Robber."[29] We both are drawn to the certainty of The Thermals' desperate, fire-and-brimstone punk in "Here's Your Future" ("It's gonna rain").[30]

Old songs we know well can take on new meaning in a climate context (the re-seeing process we discussed above). For me, Nick Drake's melancholy "Time of No Reply" evokes a whole new sense of feeling lonely in nature when considered from the perspective of solastalgia.[31] The jazzy swing of Mildred Bailey's 1940 song "There'll Be Some Changes Made" lightens the weight of climate adaptation.[32] And the nostalgia for the 1980s in Joon's remake of Bananarama's "Cruel Summer" became darkly ironic when listening amid deadly heat and wildfire.[33]

Artistic re-seeing also happens when young people encounter earlier works and reinterpret these on their own time. For James, an album by U2 and Brian Eno from before he was born felt eerily prescient. He loved discovering "The Wanderer" with vocals by Johnny Cash, about a future world "under an atomic sky" that evoked the current European climate of heat waves, war, rising nationalism, and climate refugees.

EXERCISE: Create Your Own Climate Playlist

Whether you share it with others or make it for your own personal listening, here are some tips to inspire your own climate playlist.

Consider how the song, lyrics, or melody evoke some reality of your experience with climate breakdown and the ecological threats we face, either the hard realities or moments of joy. All styles and genres are welcome, and songs need not explicitly be about climate change. In fact, it's fun when the associations are more subtle or unexpected. Trust your instincts and focus on the feelings: How do the selections help you express your climate-related emotions, what you want to feel, or what you need to feel?

One of the best things about this exercise is that everyone brings their own style. In her book *What If We Get It Right?*, for example, Ayana Elizabeth Johnson describes her climate playlist as heavy on soul, R&B, and what she affectionately calls "anthems for victory, love songs to Earth, tunes for tenacity, and sexy implementation vibes."[34] Don't fall into choosing music you think you "should" listen to. If the song doesn't vibe with you, leave it off. This isn't duty, it's expression.

DRAWING

In *The Elements of Landscape Painting*, author Suzanne Brooker writes, "If every kind of landscape—from the Grand Canyon to Mount Fuji—has already been painted, then what still makes painting a landscape interesting? The answer is you. You are the new and novel ingredient: how you see and respond to a scene, the way you translate and interpret the felt sensation of nature into a two-dimensional painted surface."[35] The boost of positive feelings you can get from coloring, drawing, or doodling happens despite any skill or training. In fact, studies have shown that even fifteen to twenty minutes of creative self-expression or art-making tasks can increase your perception of your own ideas and ability to solve problems.

WRITING

Much like other creative pastimes, free writing, journaling, or composing poetry helps you manage climate anxiety, cope, and recover from disasters, giving you an outlet for expressing your feelings, integrating memories, and becoming empowered by the act of writing itself. In psychological terms, writing creates an *internal locus of control*, a sense of personal agency over your life. Expressive writing is a proven way to work through emotional difficulties and traumas.[36]

A brief practice of free writing, inspired by Julia Cameron's book *The Artist's Way*, is a good start.[37] The method is simple: a daily habit of writing three pages, longhand, of your stream-of-consciousness thoughts, feelings, memories, or impressions first thing in the morning. Larysa found a daily free write—done at her "analog desk," away from screens—helped center her awareness and her sense of what she could control, making her much less reactive about far-off events and disaster news that used to derail her. As a new mom, she found that carving out this sliver of personal time helped her recover a sense of herself in her life post-pregnancy.

A daily creative practice can help teach self-acceptance. The poet William Stafford had a practice of doing a poem a day. When a critic asked, "What happens if a poem doesn't meet your standards?" he replied with his characteristic self-effacing wit: "I lower them." Good advice for staying the course. When Roman added a practice of journaling and crafting simple haiku poems to his sustainability pyramid, he ended up sorting through memories from all his arctic travels and research. In doing so he gained a new perspective on his doctoral program and how he might approach his dissertation and teaching.

FOUND ART

The concept of "found art" (*objet trouvé*) means art made from things that are not normally considered traditional art materials, like sculptures of threatened sea animals made from plastic litter gathered from the ocean. Found art is the beauty you discover or create in an unlikely place. Creating found art can be an antidote to the *found grief* of the climate era: the surprise emotions of loss you might feel when stumbling on a troubling news story about a looming extinction, learning that a fire or flood destroyed a place you loved, or glimpsing a dead animal on the side of the road. The unnatural disasters and weathers of climate breakdown have already shattered your sense of normal; therefore, putting natural materials into new forms and contexts helps you recover beauty and reminds you of your power to put a broken world back together.

Sam and Liberty had a practice of making nature sculptures with their kids, inspired by artists like Andy Goldsworthy: building sandcastles; arranging driftwood, shells, and feathers at the beach; and creating elaborate mandalas of small stones.[38] If they found dead animals (and deemed it safe enough), they would take the time to create a small monument, decorated with flowers and leaves, like the intricate animal memorials created by anthropologist and photographer Amanda Stronza.[39] Sometimes they made a point to dismantle their creations and scatter their stones and twigs like a Navajo sand painting or Tibetan mandala to send them back to nature (and to teach their kids "Leave No Trace" camping skills).

SUSTAINABLE CONSUMPTION

As with all your activities related to climate and environmental problems, you need to be mindful of how you engage with nature or environment-focused arts, especially as passive entertainment. The ancient Greeks had a concept, *enantiodromia*, to describe how an action begun for one reason can lead to the opposite of the intended outcome. This can happen when you impulsively choose a movie, book, or piece of music seeking escape and end up feeling even more distressed. This is an odd aspect of being a climate hostage—feeling powerless yet fascinated by problems and compelled to keep watching. For this reason, Agnes moved away from the dystopian stories she had consumed in middle school.

It is also common in our market-based society for news to be bundled with entertainment and advertising—so much so it's sometimes hard to tell the difference. Remember, despair is often fatigue in disguise. When your senses are exhausted and willpower is low, the best solution is to unplug: power off your phone or close your laptop and go outside. Like the dose-response we get when spending time in nature, the longer your breaks from technology, the more benefits to restoring your senses and sensibility.

Think in terms of your sustainability pyramid. You want a balanced diet of nourishing artistic and entertaining content: both the kind that is escapist and uplifting and the kind that will allow you to sit with difficult emotions.

You can be mindful about shadow sides of art and entertainment by keeping track of quantity, quality, and curation:

- *Quantity.* Hundreds of thousands of songs are released each day, and innumerable videos are uploaded each minute. Streaming shows release entire seasons all at once. Consider limiting the amount of time you spend scrolling through the endless options. Otherwise, keeping up with entertainment becomes another form of FOMO, stress, and fatigue.
- *Quality.* Be careful about bombarding yourself with dark material, even if that's what you seem to be craving. When seeking entertainment, check in with yourself: What are you feeling, and what do you want to be feeling? Choose accordingly.
- *Curation.* A creative sensibility can help you approach your news intake as a discerning critic, versus a passive consumer of troubling information that assaults your senses. You can ask yourself: "How am I organizing all this input for myself in a pleasing and healthy way that supports my values and best self?"

———

Add "arts and creativity" to your tool kit of strategies for coping with climate anxiety. Later, you'll get a chance to apply these ways of seeing and doing to your environmental actions. There is no one way to solve wicked ecological problems of the climate crisis, but arts challenge us to find solutions and strategies that are unique and flexible.

Chapter 16

Spirit

The earth herself, burdened and laid waste, is among the most abandoned and maltreated of our poor; she "groans in travail"... We have forgotten that we ourselves are dust of the earth...our very bodies are made up of her elements, we breathe her air and we receive life and refreshment from her waters.

—POPE FRANCIS[1]

FOR THOSE OF US WHO SUBSCRIBE TO ANY kind of spiritual or religious values or traditions, it can be difficult to reconcile our faith with the evidence of all the immense suffering climate breakdown is causing all around the world.

Cassandra, a Chinese American university chaplain, found herself struggling to respond to students who looked to her for spiritual guidance on this very question.[2] One student asked, "If God created the natural world we live in for us to care for, how can we pray to God for solutions to climate change when we ourselves are the ones who destroyed it?"[3] Another said that her spirituality resides in the woods and trees, where she had always felt awe and a sense of peace, adding that in this time of raging human-caused wildfires, her forest church had begun to feel less like a place of wonder than a place of pain. Cassandra had recently attended an interfaith climate summit, where imams, rabbis, priests, ministers, pastors, Native elders, and healers all had the same challenges on their minds.[4] Many of their congregants were in pain, with each new climate disaster fueling their existential dread and spiritual crisis. These questions deeply challenged Cassandra because she had been asking the same ones herself. A child adoptee, she did not know her background, and she often felt adrift in terms of her sense of place. She wondered how she could counsel these young people when she too was struggling for answers.

You may have faced Cassandra's dilemma. Whether you identify with a formal religious tradition, or simply believe in a higher power as a source of meaning and purpose, eco and climate problems can challenge your belief systems and complicate the spiritual dimensions of your life. At the same time, your personal spirituality and religious faith—in whatever form it takes—can be a key component of coping and even flourishing in these turbulent times.

SPIRITUALITY AND COPING

The term "religion" typically implies participation in an organized practice or shared belief, while "spirituality" is the process of seeking out the sacred, whether that is a higher power or something less easily defined. In the Pacific Northwest, where I live, about 25 percent of people select "none" in a survey about their religious affiliation, but many of the same people also identify as being deeply spiritual.[5] In fact, more than a quarter of adults in the United States say they are spiritual but not religious.[6]

Whether they are religious or not, many people consider spirituality to be a foundational part of their environmental identity and how they view the world. Moments of transcendence in the great beauty of the outdoors, epiphanies about the miracle of life, or mystical experiences of oneness and connection to nature often show up on people's eco-timelines.

Humans have long summoned their religious and spiritual resources as sources of strength during challenging times. A growing body of research in psychology and mental health therapy looks at the diverse ways people think about the sacred, and how these mindsets inform their health and well-being. Robust findings indicate that more than 70 percent of US adults consider religion to be important in their lives.[7] Studies show that religious or spiritual involvement is linked to physical health, longevity, reduced depression and anxiety, a sense of meaning and purpose, and over-all increase in life satisfaction and happiness. Climate therapy researchers like my podcast partner Panu (also a Lutheran pastor and environmental theologian at the University of Helsinki) are creating a growing body of research showing how this process can be effectively applied to coping with eco-anxiety and grief.[8]

It's easy to feel overwhelmed when living through times of existential questioning about the future of the planet. Faith in a high power—something larger than yourself, something that transcends your one and precious life—can be a guide. *Self-transcendence* can function as a strong foundation for your environmental stewardship and action here and now. It can take many forms, like the felt sense that you are part of a larger fabric of life, a larger universe story, or the realization that whatever capital-*I* goal or vision you are pursuing is just one link in a larger chain of action undertaken by many, coming before and after you.

PRAYING FOR A SOLUTION?

Should you pray for a solution to climate change? If you are a religious or spiritual person, you might easily imagine yourself seeking divine help based on your tradition and your understanding of what prayer is. Alternately, you might have a strong and even visceral negative reaction to invoking a deity to solve the physical and very human problem of climate breakdown.

Believing in the value of prayer is akin to the active practice of hope. Effort and concrete action might be needed to justify your hope, as in the proverb "Pray, but move your feet." Or maybe even some blood, sweat, and tears are required, as in the "gritty hope" evoked by ecological grief researcher Ashlee Consulo.[9] Opening yourself to sacred teachings puts you in contact with other qualities, like grace and compassion, that create a wider space for thinking about hope. Grace gives you a sense of forgiveness that doesn't need to be earned, and it's something you can grant yourself. Compassion creates room for the acceptance of conflicting feelings and moral dilemmas for both you and others.

When you unpack it, you'll likely find yourself somewhere in between on questions of faith, hope, and belief. Even the most committed atheists I know describe respect and reverence for nature and accept that people seek some kind of higher power. You may see the value of organized religion and spirituality even if you're not sure how helpful they are for you. As you gain lived experience, cope with losses and setbacks, and respond to the needs of your family and community, you'll likely find that your beliefs around spirituality and religion, much like your environmental identity, change across the seasons of your life.

MY OWN SPIRITUAL PATH: BEING A REVERENT MATERIALIST

My approach to matters of faith, spirituality, and nature certainly keeps evolving. I was raised in the Catholic Church, was influenced by the Irish and Polish working-class heritage of my parents, and attended Catholic schools. As a child, I knew what it was to pray, and to wonder in awe at the life-size statues of saints and the murals of angels above the altar in the historic Byzantine-style church in my neighborhood.[10]

In the decades since, I have sought different ways to find meaning and spiritual guidance. I'm influenced by my outdoor experiences, my travels, and my understanding of science, ecology, and psychology. These days I consider myself a "reverent materialist." I approach the world in a way that balances my scientific, empirical worldview with respect for the mystery and sacredness of the universe. I still appreciate the rituals of Catholicism and its tradition of service and good works. In daily practice, I say my higher power is "music and nature." That is where I find deeper meaning, joy, and solace.

Nature in Your Version of Spirituality or Religion

Nature and ecology can manifest in your version of spirituality or religion in multiple ways. If you imagine a spectrum of how nature is valued in organized religions, on one end you have religious teachings that place nature under the dominion of humans and God and see protection of the natural environment as a religious obligation for humans as wise stewards of creation.[11] I call this a "heaven-down" approach. Nature-centric religions and spiritual practices take a "ground-up" approach, locating God, spirit, or the sacred directly *in* nature. In this view, other species and the natural world are endowed with worth, value, and intelligence regardless of their usefulness to human beings. Rather than custodians, humans are kin to the natural world.

Rick was a longtime practitioner of transcendental meditation since he first discovered it in high school. Over the course of his life he had spent time in India and followed Buddhist religious leaders like the Dalai Lama and Thích Nhất Hạnh. For him, spirituality was integral to his sense of interbeing with nature.

For Larysa, tapping into spirituality was a challenge, as she had a secular worldview. I encouraged her to reflect on her interest in reading about astronomy and physics: a place where uplifting feelings like awe and reverence for the mysteries of the universe were natural for her. Larysa had also studied Buddhism when she was younger, and eventually found herself able to draw connections between Buddhist principles like nonattachment and her letting go of anxiety. Over time, Larysa became more comfortable with adding "spiritual practices" to her list of coping skills.

Christina found many connections between her Catholicism and burgeoning environmental identity as she undertook her Via di Francesco trek in Italy and was reminded of the call to stewardship of creation in Pope Francis's encyclical *Laudato Sí.*[12] Reid grew up in the Episcopal Church in Texas and this faith was still a large part of his family life, but he also held forms of "ground-up" beliefs that came from the spiritual sense he experienced when surfing on the Gulf Coast. As a conservation scientist who worked closely with First Nations communities in Canada, Jann found the image of "braiding sweetgrass" helpful as she sought to integrate her experiences with wild lands, her knowledge as a scientist, and Indigenous ecological wisdom.[13] And Marcus carried great respect for the role of the Black churches in anchoring African American communities, supporting civil rights, and leading the movement for environmental justice.

Rather than view these different belief styles as being in conflict, I encourage you to respect them as you can, and see them as part of a spectrum of spiritual and religious orientations that have a place in the global movement to address environmental problems and climate breakdown. To paraphrase religious scholar Karen Armstrong, it's not about what you believe; it's about how your belief translates to care and repair of the world.

EXERCISE: Spirituality and Coping

Allowing your spiritual or religious beliefs to play a greater role in your life can provide a foundation for coping with your climate anxiety. Here are some recommendations to get you started:

- Check in on how your spiritual and religious beliefs affect your mood, relationships, and health.
- Explore how religion and spirituality shape your sense of self in relation to nature and other-than-human species.
- Notice how your environmental values are related to your spiritual and religious values.
- Be aware of how eco-anxiety or grief might cause you to question your faith or spirituality.
- When struggling with issues of faith and spiritual belief, seek support from family, friends, spiritual leaders, or counselors who recognize the links between spirituality, environmental values, and mental health.

SELF-TRANSCENDENCE AND NATURE

Spending time outdoors, contemplating the cycles and rhythms of nature— your local environment, plants and animals, the sky, and the weather—is a common and reliable pathway for altering and expanding your consciousness to experience a sense of something larger than yourself. One of the biggest doorways to the sacred is the doorway you use to step outside.

Soft Fascination and Flow

A basic way that natural settings create moments of transcendence is through the quality of *soft fascination*. Appreciating a beautiful landscape or gazing up at a night sky full of stars draws your attention in an effortless way. This allows your brain's metabolism and the focusing "muscles" that you rely on to navigate through the busy and distracting world to get rest, leaving mental bandwidth open for self-reflection. The longer you experience soft fascination, the greater potential for deeper insights. A day or weekend in a beautiful place can give you the cognitive space to slow down and think through recent events. A week will give you some distance and perspective on your life. Weeks or months away in nature—like Roman's visit to Antarctica, or Christina hiking a pilgrimage route—can be life-changing.

Many of the restorative effects of nature we have been discussing throughout the book are also consciousness-expanding opportunities. These can be subtle like the sense of flow, quality of focus, or pleasurable immersion that happens on long walks, during outdoor sports like skiing or swimming, or while working in a garden. And there is life-changing awe inspired by *diminutive experiences*, in which you become aware of your insignificance in relation to the scale or grandeur of nature or the cosmos. Roman experienced this sensation in Greenland and Antarctica, where he was lucky enough to witness both the northern and southern auroras play across the dark polar skies.

Rituals

Transcendent experiences in nature can also arise from the important rituals and ceremonies of your life. Sally and Martin made their life commitment at the beach, while Braden and his wife had their wedding ceremony high at a mountain lodge. Liberty and Sam had baptisms for their children outdoors and celebrated the winter holidays with a chilly solstice bonfire each year. Jann scheduled a solo outdoor retreat each year on her birthday.

Epiphanies

Some of the more profound experiences of transcendence you may have come in the form of *environmental epiphanies*, transformative moments that shift the way we perceive our connection to nature, or our place in the web of life. Environmental psychologists Joanne Vining and Melinda Storie have catalogued such experiences. One such experience captured the life-changing impact of books such as *Silent Spring*, evoking the "eye-opening" and "life-changing" experience of waking up to the existence of pesticides "that would persist on the earth for way beyond my lifetime and do horrible things, killing indiscriminately."[14]

Another respondent, speaking about an emotional epiphany experienced in Greece while looking out at the Acropolis, described the feeling of being overcome by the idea that "our air and our mountains and our views and everything just should be there for us always."

A young woman experiencing a powerful sense of interconnection to nature describes how "a very strong feeling came over me that…I had left my human family and entered a larger family. So, I was…no longer just the daughter of my human parents; I was the daughter of the universe."

Rites of Passage: Agnes's Council of All Beings

Some of the epiphanies associated with nature and spirituality are so significant as to become *rites of passage*: events that clearly mark a transition from one stage of your life to another.

Sensing she was ready for her own rite of passage, I encouraged Agnes, now eighteen years old in the summer following high school, to get away from the city and into the woods to learn more about healthy, wild nature. She began volunteering with a local forest protection group that focused on "ground truthing"—surveying timber sales in the national forest to make sure they were legal and followed environmental guidelines.

This work earned Agnes an invitation to a special retreat for forest protectors from around the world, held several hours north in the woods of British Columbia, Canada. There she received education from scientists on wildlife corridors and regenerative forestry, from lawyers and historians about the history of environmental activism, and from Indigenous leaders on historical treaties and First Nations political programs. Agnes also completed a wilderness "un-solo" experience. The facilitators called it an "un-solo" because while Agnes spent two days camping by herself, she was not alone. She was with the place, the land and air, the other species, and her own body, mind, and spirit.

After the un-solo was over, Agnes and others came together in a *Council of All Beings*. This is a classic deep ecology ritual that had its origins in Australia and has spread worldwide.[15] Each person in the group steps away from their human identity and speaks from the place of another life form—an animal, a species, or a place. While it is a simple ritual, often powerful emotions come up on behalf of the other species: care, protectiveness, sadness, guilt, and anger. It's not uncommon for participants to cry. This experience can be a great source of epiphanies: one people remember as a crucial moment on their eco-timelines.

Agnes's Council of All Beings ceremony was elaborate and theatrical, involving a procession, handmade masks, and a nighttime gathering around a campfire. Before their un-solos, participants had each drawn the name of a forest animal to embody. Agnes drew the spider, specifically the northern orb weaver, a denizen of the old-growth forest whose large circular webs could be seen glistening with dew in the morning sun. Agnes had always viewed spiders as scary household pests. She learned that orb weavers, like the giant trees, are truly elders to humans, and have roots on this planet dating back over 140 million years. With newfound information about the spider and its role in the ecosystem, Agnes began to rethink her assumptions about these creatures and herself.

When Agnes spoke around the fire in the voice of the spider, she found herself describing how she patiently watches and sees, diligently weaving her web each night. From then on, Agnes carried a little bit of that experience with her. Later, she got a small tattoo of the orb weaver on her inner wrist to remind her of the values of patience and diligence that came to inform her environmental activism.

Psychedelic Journeying: "Trust Being"

Another inroad to spiritual awakenings with nature is through psychedelic journeying. A number of clinical trials have shown that psychedelic therapies using psilocybin mushrooms, ketamine, and MDMA are safe and effective in treating conditions like PTSD, depression, alcoholism, and anxiety about death for people enduring fatal illnesses.[16] A combination of psychedelic and talk therapy has shown benefits as well.[17]

The benefits of psilocybin as a treatment for severe climate anxiety include an increased ability to process emotions, a reduction in the brain's fear response, enhanced self-awareness, and awe-inducing feelings of unity with nature. A single experience with psilocybin is associated with increased feelings of relatedness and connection to nature.[18] In describing the epiphany he experienced on his journey with psilocybin, author Michael Pollan writes, "It's extraordinary that the plant world might be offering us an antidote to the flight from nature. These plants call us back to nature, and nothing seems more valuable right now than something with that power."[19]

James had had some casual consciousness-altering experiences in the past, with mixed results, and he wanted to try it again, but in a more intentional way. During his psychedelic journey, James was comforted by the advice his trained psychedelic medicine facilitator gave to "trust the experience": to trust the medicine—in this case the psilocybin—and the journey. James relaxed into the waves of energy and insight as his trip ebbed and flowed. During his experience, James wrote a cryptic note to himself to "trust being." Over the weeks afterward, I helped him to process this. Visualizing wading into the surf at the ocean had led to the revelation that the waves were not trying to knock him over. Whether big or small, they didn't have some agenda; they were just *being waves*. He could feel safe floating in the ocean, among the waves. It was *all being*, and he could trust this.

A hallmark of mystical psychedelic experiences is their ineffability. It's hard to put them into words; language doesn't do justice to the striking sense of knowing or discovery, or the rapid rearrangement of long-held assumptions, leaving us feeling changed in a fundamental way. In holding the space for James, I recognized that he was describing an experience of oneness or boundlessness that many others have recognized: the sense of an "oceanic feeling," popularized a hundred years ago in the writings of Sigmund Freud.[20] Back in his regular world, it took effort for James to recover this sensation. But he knew it was there, this subtly different perspective on life, death, and existence that he described as a "calm place inside me." As he later said, "I went away and I came back. I'm not sure yet what it means but I still feel it."

Touching Mortality: Liberty's Cancer Diagnosis

Many times, rites of passage are not freely chosen. This is certainly the case when you are a climate hostage, confronted with global-scale anxiety and grief. But even if you have taken on an environmental mission, life may thrust other challenges upon you.

Sam and Liberty were stunned when Liberty was diagnosed with invasive breast cancer. The shock of the diagnosis naturally prompted thoughts of worst-case scenarios, and I supported her and Sam as the family took in the reality of their situation. I drew on my behavioral medicine training and also on my personal experience of my wife's surprise diagnosis of metastatic breast cancer at age thirty-five. Luckily, Liberty's tumor was discovered early and offered good chances for successful treatment. Nevertheless, Liberty was forced to confront her mortality much sooner than she ever expected.

When sudden illness or injury strikes, the very idea of personal sustainability is cast in stark relief. Overnight, Liberty's goals went from juggling household responsibilities and making time for her advocacy work to reordering her life to prioritize her cancer treatment and being present for her kids. Meanwhile, Sam took the lead at home to give Liberty space to focus on her health and survival. Their commitment to their Families for Climate work remained, but for Liberty the actual work would have to wait. I helped her redistribute the work to others who were willing to take it on.

Just like someone weathering loss or injuries from a climate disaster, Liberty became more conscious of what it means to be a "survivor." She vowed not to take life for granted. Meanwhile, she and Sam educated themselves about how to support their kids. They learned that many of the skills and strategies they had learned to manage climate anxiety were useful in easing their children's anxiety about their mother's illness.[21] Later, when they were able to get some perspective, Sam and Liberty had more compassion for people who were disengaged from climate change because of other issues they might be coping with in their lives. They decided that if they made it through the therapy, they would devote time to helping the growing number of people suffering from cancers and other illnesses caused or exacerbated by environmental toxins and a disrupted climate.

CLIMATE BREAKDOWN AND FEAR OF DEATH

The specter of mortality and fear of death are always lurking within climate anxiety. Unfortunately, these fears will become more pronounced as global heating continues, unnatural disasters become more commonplace, and the illusion of finding a climate haven is lost or shattered. Suppressing fear of death can be one of the biggest barriers to addressing climate anxiety if it keeps you mired in a sense of helplessness and dread. When you accept that, even in good times, your tenure on Earth is brief and precious, this reality can liberate you. To the extent you can transcend the fear of death underlying climate anxiety, you free up that energy for life.

Everyone gets to do their own apocalypse. Or, as we discussed in Chapter 11, people grieve in character. Though it is uncomfortable, confronting your mortality can be a powerful way to reorient your climate concerns. Liberty's cancer diagnosis prompted her and Sam to review their priorities and get more clear about what they hoped to achieve as individuals and as a family. And when I had a candid talk about death with Larysa, it helped her move past her obsessions with disaster prepping.

I use an advanced exposure therapy visualization to bring people directly into—and through—their existential fears of ecological destruction (and, implicit in these, their own mortality). It builds on concepts like anticipatory grief we discussed earlier. I developed it as a sort of stress test for the first group of counseling students who completed the ecopsychology program I created. Could they remain present with their own fears so that they could be present for others? We called it "The Bulldozer Exercise." It helps us remember that we are here in our physical forms for a short time, while life on Earth endures. (Use your judgment about the level of emotional challenge it brings for you, and feel free to take a break from the exercise and come back to it if you need to.)

The Bulldozer Exercise

Breathe and relax your body. And make sure to keep your breathing steady and deep. Imagine you're deep in the middle of a forest, maybe a rainforest. It's late morning with a clear sky. You feel the heat and humidity under the shade of the tree canopy, and you're attuned to the deep silences and the sounds of life: the ripples of wind, bird calls, activity of small mammals, buzzing of insects, waters flowing. You notice something at the edge of your awareness. An odd silence has overtaken the forest. You observe that the animals are moving differently. Large flocks of birds fly overhead. Soon you make out a distant rumble of something mechanical. And you start to feel vibrations from the ground. You breathe.

It's clear now that the animals around you are agitated and are moving away from the source of the disturbance. Soon you clearly hear the roaring of engines. You can see the tops of trees shaking in the distance, and hear tree trunks falling to the ground. The ground beneath your feet begins to shudder and quake. Small animals run past your feet. Finally, you can make out the row of large bulldozers moving through the forest, knocking down everything in their path. As you stay rooted to your place, the machines come closer. You can see the large steel earthmoving blades and the human drivers high up at the controls. The air is thick with dust, diesel fumes, and the sharp smell of torn vegetation and turned-up earth. You breathe and remain present.

Soon a scuffed blade is looming over you. You are enveloped in darkness and knocked down, along with the tree trunks, branches, roots, leaves, and stones. Eventually the machines pass overhead and move off into the distance. The noise and shaking begin to abate and soon it is silent again. The razed forest lies baking under the midday sun.

You breathe. An unknown time passes.

And then all at once, you become aware of life. You sense the stirring of insects and small animals burrowing through the soil around you. You feel trees *re-rooting* and new plants taking hold. Eventually you hear birds singing. Days, nights, and seasons pass. The forest begins to regenerate, to regrow. And you reawaken along with it.

The Bulldozer Exercise is a thought experiment about the death and rebirth of nature: a way of going deep inside your fears of ecological destruction, letting your nightmares play out to the bitter end, remaining conscious, and realizing at the end that you are still here. You still have choices. You still have possibilities. You can replay the exercise and adjust the scenario, keeping in mind that this is your imagination, you are choosing to face your fears, and you can breathe and keep your body calm. Perhaps you choose to allow a hurricane to sweep over you: the arrival of the winds and rain, the storm surge, the swirl of the cyclone, the brief calm as the eye of the storm passes, and then the return of the maelstrom and the flooded aftermath. Or you move through the cycle of a wildfire and the regeneration of a forest or a community. Or imagine enduring a heat wave.

After doing these visualizations, people are often silent, sitting with the experience. Sometimes there's confusion. What just happened? Then a sense of emotional relief, some sort of catharsis or letting go, or lessening of a weight or burden. It's one thing to understand intellectually that a forest will regenerate itself, that life is tenacious and will find a way. It's another thing to imagine yourself living through it, to be able to say, "I endured destruction and I am still here."

YOU'RE NOT DEAD YET

Even as ecological crises compel you to contemplate death, you still have a lot of living left to do. One of the many provocative ideas from the Jungian analyst James Hillman is his "acorn theory" of human development, which argues that we best appreciate a life by looking at it from its end, what it has turned out to be. You are not yet at the end.

Like art and creativity, your religious or spiritual beliefs inform how you see the world and yourself in it. Spiritual contemplation sets the stage for conscious action, and can be an action itself. But for now, take a moment to reflect on how spirituality and religious belief may help you with the big existential questions about life, death, and the future of the planet.

PART FIVE

Action

Anyone reading this, particularly if you were born after the 1960s, has been surrounded by messages about protecting the environment for as long as they can remember. Each generation has its own eco-zeitgeist and its most pressing issues. I remember decorating my 1970s grade-school notebooks with "Save the Bald Eagle" stickers, along with space-age doodles about astronauts.[1] But at a time when so many environmental issues feel pressing, it can be hard to know which ones to devote your time, attention, and action to. Without preparation, getting involved in environmental action can feel like heading onto a playing field when you don't know what sport is being played, what your position is, what equipment you have, or what league you are in.

Luckily, you have been building the necessary foundation. The process began in Part One, with awareness and intention regarding how you think about wicked eco and climate problems, and developing a growth mindset about your abilities to respond. You added emotional intelligence—understanding what you're feeling, reaching for what you want to feel, and having the wisdom and capacity to stay with feelings, even uncomfortable ones, when this is the path for honesty and growth. You stopped to take a breath and reminded yourself that, in terms of eco-stresses, "the body keeps the score," and attending to your personal health is itself a form of sustainability and environmental action. You began to take a clear look at your own "Individual Problems with Climate Change" report in the form of threats to your family and community, and to prioritize the adaptation efforts you need to stay healthy amid the unnatural weathers of a changed world.

Larysa was able to honor her climate anxiety as a new mother and move from a place of late-night doomscrolling to hands-on adaptation with her neighborhood disaster response volunteers. Marcus learned to take better care of his body while he fought for climate justice, and left his work scowl at the door so he could be a nurturing dad. Delphina found ways to overcome her summer depression so she could continue her needed work studying "forever chemicals."

In Part Two you learned that your efforts on behalf of nature and Earth can flow from your environmental identity and what you love and value about the natural world. You revealed a spectrum of nature settings where you can find your niche. You discovered your family tree is also an eco–family tree that holds stories about nature and environmental identity over generations.

Jann found she could trace her career in wildlife conservation on a timeline going back to her childhood. Reid made choices to protect his young family and maintain connections with his home place and the people he loved. Christina faced up to her sense of disconnection and inadequacy about doing outdoor activities, started a hiking practice, and expanded her comfort zone to include an international pilgrimage.

In Part Three, you took up healing practices for the worst mental health impacts of eco and climate issues: severe anxiety, despair, a lost sense of safety outdoors, and the trauma of disasters. You learned to give your environmental identity a standing place, or places, that you can learn about and commit to protect. Care of nature includes nurturing damaged landscapes that need our love.

Jesse slowly recovered from the loss of her family forest land to a catastrophic wildfire, and reinvested herself in life and relationships. Roman, the Antarctic researcher, worked through his inner turmoil about the impending loss of Earth's ice caps and his guilt about his privileges to travel, study, and become a father. He practiced being open and vulnerable about his feelings as he and his fiancée sorted out their life plans. Rick was able to mourn the losses he experienced in his long career as an environmentalist and share his wisdom with young folks who were continuing the tradition.

In Part Four you experimented with ethical happiness and thriving. You learned how your ecological self is anchored in and nourished by your relationships, your creative sensibility, and spiritual or religious values.

Slade discovered how to calm his nervous system and lessen his mood swings through the power of music. Rachel became empowered to speak out as a mother and mental health professional once she made peace with unresolved issues in her family. James experienced the transcendence of being at one with nature on a guided psilocybin journey.

Now we'll move to taking action: the final part of the cycle of surviving climate anxiety. Remember that anxiety is simply a healthy signal to investigate a threat. By taking action against that threat, you discharge the energy of anxiety built up in your body and clear the way for new feelings and information to come.

In the chapters ahead, I'll show how engaging in climate action can open up a world of opportunities to lean into your eco-values, gain a sense of control and agency, and find productive outlets for your difficult feelings. I'll help you vet different types of action, identify a mission, and create a style of going after it that flows from your values and your life context. And I'll give you some roles to take as you navigate the challenges and barriers that await you when you try to make positive change. Just as there is a spectrum of nature settings for you to connect with and protect, there is a wide spectrum of goals to commit to, and strategies for getting there. The chapters ahead will help you find the ones that are right for you.

Chapter 17

Duty

It's better to do your own duty, even poorly, than another's duty well.

—Bhagavad Gita

I'm doing a climate workshop with university students, faculty, staff, and members of the local community. We've been discussing environmental identity and ways to understand everyone's values and special connections with nature. Now participants are looking for ways to take action. I ask everyone to stand. I point to the front of the room and say, "In this direction, imagine this is the front line of climate action: the realm of nonviolent protest, or civil disobedience, where it is tense, loud, and potentially dangerous. It might be a pipeline protest, or where people sit in old-growth trees to protect them from being cut down, or a hunger strike to protest government inaction on climate deaths and disasters." Then I turn, point to the back, and say, "In this direction, you can stand as far back behind the front line as you want." Finally I instruct the group to line up, in one single line, from the front to the back.

But of course it's not easy for people to make a decision about how close to the action they want to stand. On a deeper level, the question is one of duty: "What is expected of me? What do I expect of myself?"[1] As often happens, some people are immediately decisive, moving to a spot in the front, middle, or back. But most of the group hesitate. They want more information: "What's the issue at hand? What is the level of danger? What if I'm not sure?"

I gently encourage them to do the best they can with the information they have. People mill around and confer. Some take a place toward the front. Others filter toward the middle. Some slide toward the back. Eventually everyone settles into one line. Then I have them look around and see where they ended up.

At the very front, Keysha, a young biracial woman, explains that she is an "intersectional environmentalist" committed to social justice for women and people of color, but also for the planet and other-than-human species. Evan, a tall young white man, says he's at the front because he wants to help organize and protect people. Toward the middle, Mai, a professor, says her skills would be better suited for supporting frontline action through her community organizing and advocacy work, and, frankly, she's afraid of potential violence at the front. Near the back of the line, Abbie, a graduate student who is showing her midterm pregnancy, says that with a baby on the way she feels more comfortable supporting from behind the scenes. Klara, a visiting scholar from Ukraine, moves to the back of the line. Life has been difficult. She is going to take advantage of the opportunity to rest.

I know a professor who teaches about civil disobedience. On the first day of class, he asks his students, "What would you get arrested for?" As with the answer to that question, no place in the line is right or wrong. It's about becoming conscious of your comfort zone and appetite for taking action. Most people suffering from climate anxiety harbor guilt or insecurity that they are not doing enough and may implicitly see the purest form of environmental action as direct protest, putting their bodies on the line.

Make no mistake, we need people on the front line, and that might be your role. "Power concedes nothing without a demand," as American abolitionist Frederick Douglass succinctly observed.[2] But as much as we need people who are committed to direct action, there are many important ways to be an agent of change. Those at the front line require support in the form of aid and encouragement. We need people working behind the scenes to craft strategy and messaging, work on policy and change within the system, invent new technologies, grow food, take care of children and elders, and make art to inspire people. The takeaway is that there is a place for everyone. Your sense of duty (literally, what is due, what you owe)—to Earth, your community, your family, or yourself—is yours to determine. Remember, you *are* nature. You are part of the ecosystem. Like Liberty coping with breast cancer, there are times when sustainability means taking care of yourself, resting, and healing. You may simply bear witness to problems as honestly as you can and push against denial—there is honor and integrity in that.

Moreover, this exercise highlights the problematic assumption that you have a *choice* about whether to take action. The decision to hang back or sit out is a privilege that people take for granted. Some have the luxury of living their life as if climate change doesn't exist. But if you're a member of a historically marginalized group or have felt the brunt of climate injustices, you know some people don't have a choice, that they are born on the front line.

Sometimes you may find yourself thrust into action when an issue directly affects your life or threatens something you love and care about. If I name different scenarios—protecting your family from harmful chemical exposures, standing against violence toward peaceful protesters, or stopping destruction of a wild landscape—this affects your urge to take a frontline position. And if I ask you to imagine how you'd react at other times in your life, the calculus changes. When I was a young man, I would have been like Evan, ready to be in the vanguard for almost any worthy issue. But as a parent, with more specialized expertise to offer, I am more like Mai and selective about my actions and how I manage risk.

The advice from the Bhagavad Gita is to do your own duty, even if it is imperfect, flawed, incomplete, or a fledgling effort.[3] Your path is your own to seek. Even if it takes time and experimentation to find your standing place, once there, you're home.

AGENCY WITHIN YOUR AGENCY

If your agency is your capacity to exert your power through action, revising your relationship to frontline action gives you *agency within your agency*. It gives you the power to choose how and when to use your power.

For James, an activist in his mid-twenties, this exercise sparked an epiphany. He had unconsciously assumed he *always* needed to be at the front of the line. He began to realize that over the course of his life he might sometimes be near the front and other times could take a break or position himself in other ways. The realization not only freed him from a pervasive sense of guilt but also opened up new opportunities to offer his support.

An elder like Rick could look back and see how he had occupied a variety of places on the spectrum of action: from youthful protestor to educator, legal advocate, and sustainability theorist, and then to someone who stepped away from the field of action to pursue a spiritual, monastic path. Now he was revisiting advocacy through his work with the Citizens' Climate Lobby and meeting with the young folks in the Sunrise Movement. And Braden, who had never considered himself an activist or protestor, saw that from within the timber industry, he could advocate for change in ways that did not conform to an activist stereotype.

All three used their agency to choose their own path of eco-action. You too, can decide when and how to use your power.

LAYING OUT THE CARDS: FIGURING OUT WHAT MOTIVATES YOU

When I met Larysa, she knew her obsessive news intake was unhealthy but felt a strong duty to be aware of looming disasters in order to protect her children. She felt stuck. When you have the disconcerting experience of several conflicting feelings arising, it's natural to struggle to find motivation. The first step is to be honest about your fear and ambivalence. Earlier, in Chapter 3, you did a self-assessment to identify the bricks or foundational self-care actions for your personal sustainability. Now it's time to use a similar approach to sort out your motivations.

When you're stuck, take stock of all the different options and contradictions. I call it "laying out the cards," with each card representing a different factor in your decision-making process.[4] For Larysa, one card represented news about climate change and disasters. On one hand, she was drawn to it for survival reasons. But she was also repelled by how compulsive, unhealthy, and counterproductive it felt, like an addiction.

Another card represented her options, which upon first inspection seemed unclear. She wanted to be an engaged person, but her confidence in her ability to do so in a balanced way was low. Why was this? Typically Larysa felt confident in her life. Her goal to be a good mom was clear. When she sat with all of the different aspects of the situation, Larysa realized that her news watching had reached its logical conclusion. Recycling the same information that she "already knew" simply heightened her emotions or anxiety. To free herself from this cycle, she realized she needed to do something else with the information. This was a breakthrough that ultimately led her to join her local Community Emergency Response Team. Now, instead of worrying or ruminating about the information she was consuming, she could put it to work.

EXERCISE: The Dilemmas of a Climate Hostage

If you're not sure what actions to take, you can start by naming your conflicting motivations, what psychologists call *approach / avoid dilemmas*. Exploring the things you want (what you want to approach) and the things you are trying to get away from (what you want to avoid) gets you one step closer to identifying the actions that will make sense for you.

Reid wanted to relocate his young family to a safer place, away from worsening disaster risks—but knew it would be painful to leave the home he loved and all his relatives and friends. Rachel wanted to live a low-carbon lifestyle—but also wanted to support her daughter's cross-country journey to college and travel to see her, despite the carbon pollution this would cause. And Delphina wanted to commit to a life with her partner without the expectation of having children—but she also dreaded disappointing her parents and abandoning family traditions.

What about you?

A PATHWAY FROM VALUES TO ACTION

What you may have noticed is that when you begin to clarify your values and beliefs, move through your anxieties and despair, and begin to lay out your options, the impulse to take action can arise on its own. You start to feel ready, or at least ready enough. The next step is to channel this energy toward your own personal action style.[5]

What does it look like to have your actions flow from your environmental identity and values? Professionals in public service and conservation, like Marcus and Jann, found a clear path from their basic environmental values to their personal style of engagement. But for many others, it may take some self-reflection to connect their values to the right activities and sources of motivation.

To find your values-to-action pathway, ask yourself a series of questions, based on the material we have been covering in the book:

1. What are your underlying environmental values?
2. What are your beliefs about nature and ecology?
3. Considering personal sustainability and your "Individual Problems with Climate Change" report, what things do you value that are under threat?
4. Looking back on your eco-timeline and your unique life path, what are your capabilities to help?
5. In terms of your relationships with people and nature: What is your sense of duty and obligation? How are these inspired by your creativity and spirituality, and how can they play a role in your healing?
6. In thinking about how close you want to be to the front line of action, what actions and behaviors make sense for you?

Remember, your values interact with your beliefs about the world, as well as your sense of threat. If you believe you can make a difference in some way, your values will motivate you to act.[6] The specific actions you should take depend on your experience, location, style, preferences, and opportunities.

Marcus's strongest eco-values were human-centric, focusing on the marginalized people in his community. He had developed an ecological worldview, seeing the connections between environmental issues and health. There was a clear sense of threat to people and communities he valued, in the form of racism and environmental injustice. And he perceived that he could make a difference as an urban planner. Marcus felt an obligation to act and had pursued a career in local government.

Jann had developed strong altruistic values regarding the well-being of other-than-human species, her ecological view focused on wildlife biology and protecting ecosystems. Like Marcus, she saw clear threats, in her case to the many species in the national parks of the Canadian Rockies. For many years she found a source of efficacy in her research and administrative roles. More recently, she sought a career change and was searching for her next project. Therefore, her action pathway was evolving.

Rachel, who was not an environmental professional, needed help to discern her values-to-action pathway. She had already plunged into action with a group of other newly climate-conscious therapists. I helped Rachel fill in the gaps and build a stronger foundation for long-term action by reflecting on her environmental identity and the earlier life experiences that led her to that moment. She was able to look back on her childhood and the values

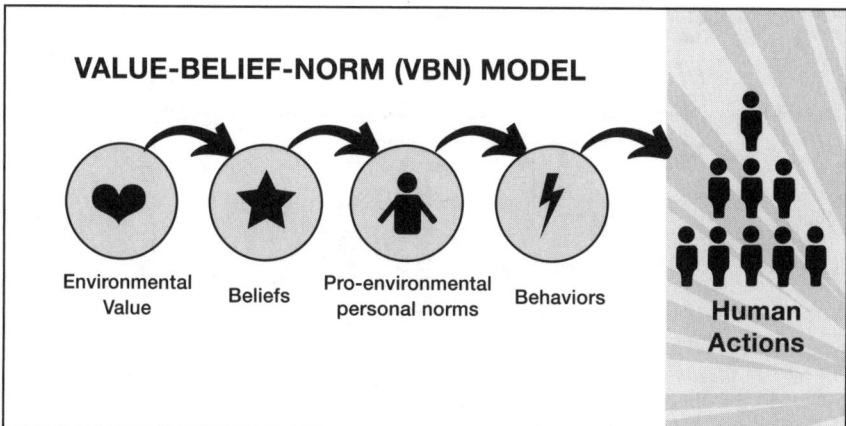

VALUE-BELIEF-NORM (VBN) MODEL

Environmental Value → Beliefs → Pro-environmental personal norms → Behaviors → **Human Actions**

Values, Beliefs, Norms, and Environmental Action

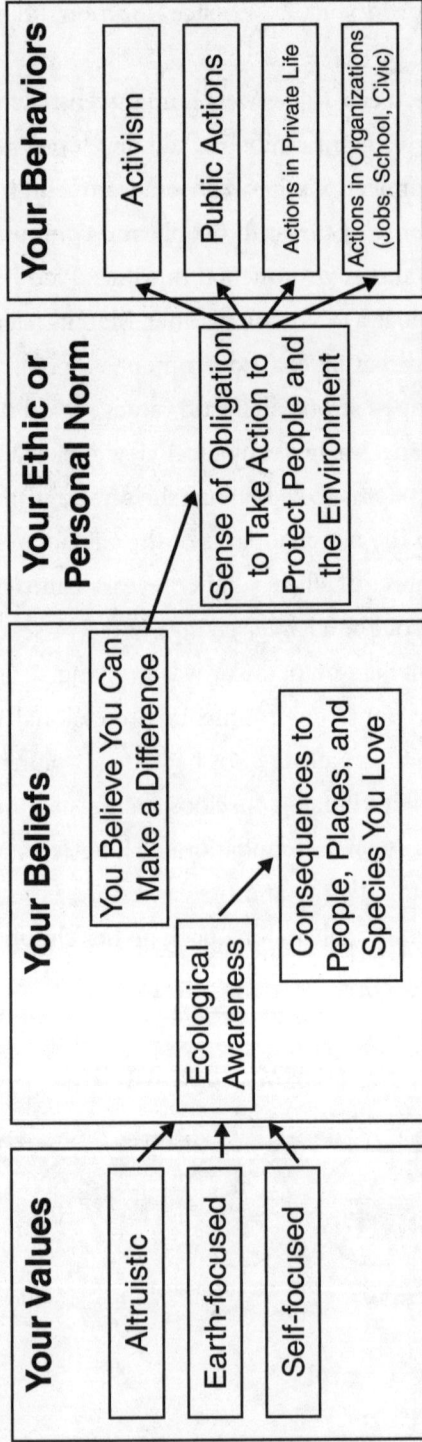

Adapted from Stern, P. C. (2000). Toward a coherent theory of environmentally significant behavior.

she had learned from her parents regarding nature and social justice. Rachel had a long-standing affinity for the concept of *tikkun olam*, an ideal in Judaism about humans' duty to repair or mend the world, which she had learned from her parents.[7] With her newfound awareness, this sense of duty took the form of learning about climate psychology, with the goal of integrating it into her therapy work to help people cope with climate distress.

LESSONS FROM LONG-TERM ACTIVISTS

In my experience, the most common barrier to eco-action is feeling inadequate to the task. Even when you are motivated, you may not be sure what you can do, or you may believe that your actions will not make a difference. We all find ourselves here. Even experts and veteran environmental advocates had to start somewhere.

For years Christina had been actively working with the Citizens' Climate Lobby, meeting regularly with her state legislators and congresspeople to promote a transition to renewable energy.[8] There had been some progress, and many setbacks. Recently, some lawmakers had succeeded in rolling back her state's involvement in a regional initiative to limit greenhouse gas emissions, an involvement that Christina and her group had lobbied for years to achieve. It was easy to get discouraged.

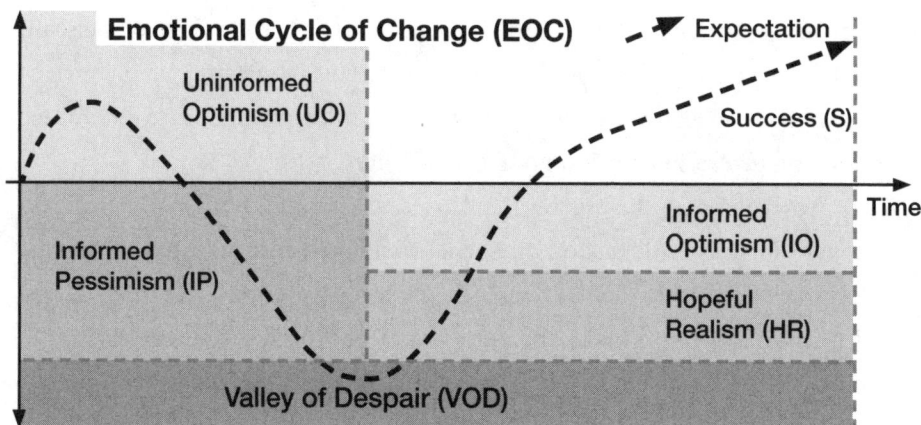

Emotional Cycle of Change (EOC) — Expectation

Uninformed Optimism (UO)

Success (S)

Informed Pessimism (IP)

Informed Optimism (IO)

Time

Hopeful Realism (HR)

Valley of Despair (VOD)

Within the emotional cycles of change, there is a despair and empowerment curve.[9] When taking on any eco-challenge, your initial excitement and enthusiasm will inevitably give way to a more sobering view of the difficulties involved. Periods of despair need to be weathered until you regain the optimism and energy to take on new goals.

When Christina came to me feeling discouraged, I helped her reflect on her political journey. To do this, I told her about a study from sociologists Willett Kempton and Dorothy Holland that found three independent factors that contributed to the identity development of long-term environmental activists and had helped them sustain action over time:[10]

1. *Salience*, a sense of waking up or becoming aware that led to a sense of environmental problems being more prominent or acute.
2. Having the *identity* of someone who takes action, and the empowerment created by taking on roles and responsibilities.
3. Having existing expertise or gaining *practical knowledge* by taking action on an issue or problem, to the point where you can teach or guide others.

The research shows us that there are plenty of pathways to finding your place on the spectrum of action, and multiple paths to becoming a leader and change agent. You might have expertise and be called on to take the lead, like many climate scientists who become public educators and policy advocates. You might have a wake-up call about a problem that compels you to take the lead; maybe you discover toxic chemicals affecting your community.[11] You can also be drawn in simply by volunteering to be on a committee or being placed in a position of responsibility.

You can begin the process like Larysa as a parent, Braden as a business owner, Reid as an educator, or Rachel as a healthcare provider. Whatever the entry point, once you are engaged, the three factors operate in a synergy. The more actions you take, the more practical experience you gain, and the more aware you become. The increased salience of problems—and potential solutions—leads to more possibilities for action and experience. This identity process continues regardless of whether your initial efforts succeed or fail. In fact, it's realistic to expect failure or setbacks when you take on new complex challenges. So *the despair and empowerment curve is also a learning curve*. With practice you can trust in your ability to take action and learn, leading you to find security in your identity as an empowered person.

The Despair and Empowerment Curve

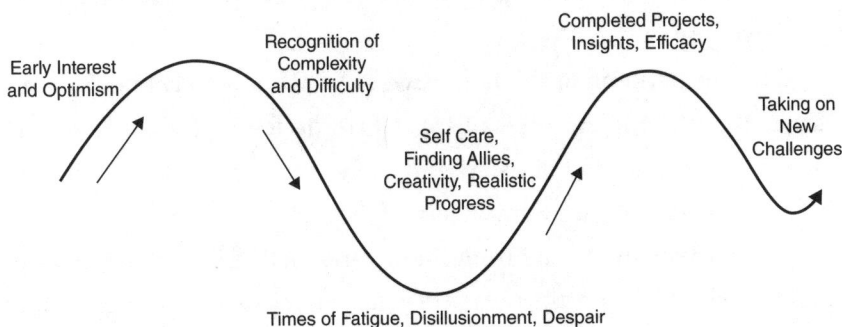

Early Interest and Optimism

Recognition of Complexity and Difficulty

Completed Projects, Insights, Efficacy

Taking on New Challenges

Self Care, Finding Allies, Creativity, Realistic Progress

Times of Fatigue, Disillusionment, Despair

We can add one more factor that helps sustain your long-term motivation: being part of a like-minded group or community. Rarely is environmental action undertaken alone, and the shared experience can fuel a collective sense of efficacy, motivation, and empowerment. Considering these internal factors was helpful for Christina. It allowed her to look back on her first experiences of environmental awareness and her tentative attempts at action and recognize how much knowledge and expertise she had acquired about state politics and lobbying. She realized that her recent policy setbacks, while troubling and disheartening, had provided valuable learnings, and this in itself was motivating. Christina was not naive. She knew that politics was a long game. But having a way to map her progress and cope with the emotions made it easier to stay the course.

HIKE YOUR OWN HIKE

Hikers on the Appalachian Trail have a motto: "Hike your own hike." Christina picked this up in her pilgrimage training, and it allowed her to be content with her own pace when she was starting out. She didn't have to keep up with the faster hikers, nor did she have to hang back with a slower group. She started using this motto in her political work too, allowing herself to focus on what she thought was important in the long term, even if it wasn't blowing with the current political winds. She returned to it again during her Via di Francesco journey, when she was overwhelmed by all the new sights and sounds and the culture of Italy. She allowed herself to express her spirituality and her appreciation for nature and the landscape in her own way.

Late one afternoon at the hermitage of St. Francis on Monte Subasio, near Assisi, after the crowds had left, Christina felt a profound sense of peace amid the quiet and birdsong and had the epiphany that her pilgrimage and her political work were all part of the same path.

Whether you are on the front lines, participating in direct action, or hanging back and supporting where you can, taking care of yourself is both a duty and a political act.

Chapter 18

Challenges

I walk down the street.
There is a deep hole in the sidewalk.
I fall in.

—Portia Nelson[1]

As a psychologist who helps people do the hard work of personal growth and change, I am exquisitely attuned to barriers. In fact, often my work doesn't really begin until someone meets a barrier. As I tell clients, if the problem were easy to solve, you would have already solved it. This is especially true with climate anxiety, given the range of forces working against you.

Cassandra, the university chaplain we met earlier, was troubled by a sense she wasn't doing enough to make a difference about environmental problems in her life or setting enough of an example for her students or congregation. Like many people, she had heard about the *personal carbon footprint* and used it as a lens to gauge her efforts. This was a good start; it was concrete and actionable. But focusing on her personal footprint led to an inordinate focus on the effects of her own actions rather than the larger political and economic drivers of climate problems. This limited her ability to have a healthy perspective and gain satisfaction from the efforts she did make. She came to realize one of the insidious things about climate propaganda is that it makes you feel responsible for things you did not cause and can't really control.

FACING THE HARD TRUTHS

After working with people of all ages on ways to express and embody their environmental values, I've learned there is a responsibility to be honest about barriers, whether you're talking to innocent children or elders facing mortality. Standing between a safe and healthy world and the many solutions to runaway climate breakdown are a number of obstacles. Many are personal, but some, as we will see in this chapter, are also political.

It's been said that "the truth will set you free, but first it will make you miserable."[2] In this chapter, we will face some hard truths. The maturity and wisdom here, or "climate adulting," is the recognition that it's not your fault. You can do your part, but many of the barriers are not about your individual actions. In most cases, *climate change is caused not by what you are doing* but by barriers that prevent societies from self-correcting by employing policies and technologies that can slow, stop, and reverse climate breakdown. The resulting inertia contributes to our climate hostage feelings. The biggest barriers are structural, economic, and political. The lesson for Cassandra and for all of us as individuals is that yes, we can take action to express our environmental values and identity. But we need to understand that our small actions will be effective only if they become part of larger-scale progress.

In this chapter we are going to bear witness to a number of problematic barriers that contribute to environmental injustice and oppression, stand in the way of societal progress toward a safer, more healthy use of planetary resources, and can impair your own coping and personal sustainability. I'll ask you to select several to keep in mind and anticipate when you take action. This will expand your capacity to take in the breadth of the issues and be a better-informed eco-citizen. When you endeavor to make social or political change or take on personal goals, you'll be aware of what you are getting into.

General Troubleshooting Advice for When You Encounter Barriers to Eco and Climate Action

- Expect you will meet obstacles, and roll with the challenges they present.
- When possible, anticipate barriers and factor them into your planning.
- Don't fight every battle. Isolate immovable barriers and go around them (an "island-hopping strategy").
- Unravel the knot: Break the challenge into smaller parts and address each from easiest to hardest.
- Wear down obstacles over time through continual action.
- Accept that some wicked problems are intractable. Instead of trying to eliminate them, work on reducing the harm they cause.
- When your survival is threatened, use bold and provocative action to draw attention to issues and injustice, engage in nonviolent protest, and apply humane counterforce to oppressive acts.[3]

MEETING THE CHARACTERS: THE DETECTIVE, THE CHAMPION, THE SURVIVOR

Cassandra was a natural storyteller. To tap into her love of narrative, I offered some character roles she could encourage her students and congregants to play as they investigated injustices, battled societal inertia, and worked on being their best self.

- *The climate detective.* Your ethics impel you to expose wrong-doing and seek justice. You use your investigational skills and pursue legal strategies to hold bad actors to account.
- *The climate champion.* Working toward reform and positive social change is your calling. You step up and champion the cause. You pursue action through government, through civil society organizations, and at the grassroots and family levels.
- *The climate survivor.* Your duty begins with a focus on personal survival and well-being. You find a place to dig in, and take action from there.

These roles are not mutually exclusive. You can embody more than one simultaneously, or in different realms of your life. For example, as in her work as teacher and clergy member, Cassandra could help others challenge unjust systems by taking a social justice approach to their spirituality. And she could also channel her own ecological values by joining a local movement lobbying for climate policy reform.

THE CLIMATE DETECTIVE: BATTLING FORCES THAT STIFLE A HEALTHY SOCIETAL RESPONSE

Imagine that you are a sleuth in a detective story, investigating a crime scene. There are many clues. There are numerous suspects. And there are countless fatalities. Sadly, that is not an inappropriate metaphor in the era of climate breakdown. There are bodies, decimated buildings and communities, dead pets and wild animals, wounded bystanders, children orphaned or made homeless. There is an expectation that the death and destruction will continue, that the "climate killer" will strike again. And the demand to hold the perpetrators accountable grows stronger by the day.

This detective archetype is found in stories from all around the world. Detectives can be amateur problem-solvers, professional investigators, official law enforcers, and forensics experts. Journalists and activists join in too. There are often cover-ups, red herrings, and conspiracies. All these plot points are part of the story of climate breakdown too.

Part of waking up to climate breakdown and other environmental issues is becoming aware of the unsavory forces that allow the perpetrators to evade accountability. If you spend time investigating the *climate crime scene*, here are some barriers you're likely to discover.

The Climate Change Countermovement

Sociologist Robert Brulle described the coalition that works to deny, obscure, and minimize the dangers of global heating as the *climate change countermovement*.[4] This group pools resources from a large number of corporations and industries, along with allied trade associations and political groups, and creates sophisticated political campaigns designed to oppose efforts to address climate change.

Corruption

It is said that the price of liberty is eternal vigilance. Fossil fuel companies bury their own damaging findings. The largest and most respected auto companies cheat on emissions. Unfortunately, when it comes to environmental laws, there is no loophole that won't be exploited. The key to fighting corruption is recognizing when the game is rigged, and stepping up when action and enforcement are needed.

Durable Inequalities

It's all too easy to feel helpless or even apathetic when faced with the realities that similar social, economic, and environmental inequalities continue to be re-created (in neighborhoods, cities, and societies), and that certain groups remain marginalized despite economic growth and technological innovations. These inequalities are not inevitable—and their stubborn persistence is all the more reason to meet them with stubborn resistance.

Fossil Fuel Subsidies

Conservative estimates place US subsidies to the fossil fuel industry at roughly $20 billion per year and European Union subsidies to be about €55 billion annually.[5] Fossil fuel subsidies are outdated at a time when renewable energy technology is increasingly cost-effective and needed to mitigate climate change, yet these subsidies remain embedded within politics and tax codes.

Profit Above All Else

The prioritization of *profit above all else*—what journalist Thomas Friedman has called the "golden straitjacket" of globalization—chains people and nations to the pursuit of perpetual economic growth. This perpetual growth model contributes to exponentially rising carbon emissions, global climate breakdown, and its unequal harms.

Psychopathology

Barriers to action on climate change can spring from the personality of leaders in government and business. In times of crisis and upheaval, charismatic leaders with strong self-belief and unwavering confidence in their positions can provide a sense of security. But when leaders demonstrate the so-called dark traits of narcissism (excessive self-importance and grandiosity), psychopathy (deceitfulness and lack of empathy), and Machiavellianism (manipulative behavior), they can compromise the performance of organizations and threaten the public good.[6] Work for robust democracy and corporate oversight.

Religious Fundamentalism

Religious fundamentalism can be a barrier to change when it denies objective scientific observations and curtails free thought and action. This can be particularly destructive when religious power aligns with oppressive governments or extractive industries. Fortunately, many aspects of religion are compatible with and even motivate environmental stewardship and climate action.

Ulterior Motives

As Upton Sinclair once put it, "It is difficult to get a man to understand something, when his salary depends on his not understanding it."[7] Some people are in a bind because their means of making a living is tied to acts that harm the environment, such as employees of companies in extractive industries. On a macro scale, people in power can have hidden financial agendas, as with legislators or judges who receive donations from the fossil fuel industry.[8]

Unregulated Capitalism

Capitalism itself is not a barrier to climate action. Market-based initiatives can solve environmental problems when employed intentionally, as with cap-and-trade programs that reduced coal plant emissions causing acid rain.[9] However, we cannot entrust the greater good of humans and the planet to an invisible hand. When employing free market strategies, we need to implement guardrails to ensure that making a profit doesn't take precedence over safety and well-being.

Violence

The fear of injury or death is a significant barrier to eco-action. Unfortunately, people all over the world who speak out against the harm caused by climate breakdown, deforestation, pollution, and illegal land grabs face threats of retaliation and violence. Nearly 200 environmental activists are murdered each year.[10]

War

The most destructive environmental stressor on planet Earth is war.[11] The largest fossil fuel users and protectors of the fossil fuel system are the combined global militaries. To work for peace and demilitarization is to work toward environmental protection and sustainability.

Wealth Blindness

Wealth in itself is not inherently bad or evil. But it can insulate us from and blind us to many of the world's problems. Moreover, once you are habituated to a certain standard of living, it is easy to become entitled and unwilling to give it up. Despite being destabilizing and unhealthy for societies, systems that perpetuate extreme economic inequality are often rationalized as inevitable, when inequality is perpetuated and exacerbated by laws and economic policies.[12] These we have the power to change.

THE CLIMATE CHAMPION:
SUPPORTING HEALTHY SOCIETAL RESPONSES

Every project needs a champion. The timeless story structure known as the hero's journey is a fitting model for eco and climate action. First you refuse the call. Any reasonable person tries to avoid embarking on a perilous and uncertain quest. But then there is the disaster that impels you to the front line of action. Wildfire or floods ravage your home place and you become a climate refugee, for example. Then there are challenges and temptations: learning to negotiate wicked problems, dealing with losses and reversals, resisting despair, and freeing yourself of false teachings and propaganda. Along the way you find like-minded allies, including trusted mentors who equip you with new knowledge to aid in your journey. Eventually you return to your community, transformed, bearing a gift or treasure that helps solve some aspect of the climate crisis. Others look to you as a role model and inspiration. Climate and environmental issues occur worldwide, so many champions are needed.

You will encounter dragons—in this case, the barriers to environmental action that Canadian environmental psychologist Robert Gifford dubbed the "dragons of inaction."[13] It's good to remind yourself of these common obstacles, especially as many remain somewhat hidden.

"Black Swans"

A "black swan" is a metaphor popularized by statistician and author Nassim Nicholas Taleb that refers to highly improbable, rare, or unforeseen events that have serious consequences. Viewing climate disasters as black swans is a cognitive error that can dampen motivation. The reality is that climate disasters are neither rare nor unforeseen in today's world.

Code Debt

This concept from computer programming refers to the problematic reliance on outmoded software, equipment, or technology, which begins to require that more and more time and energy be spent on patches and fixes rather than the original function.[14] On a personal level, code debt is a metaphor for dependence on outmoded patterns of thinking and acting, like fixed mindsets, that no longer serve you. The problems of climate change are a kind of societal code debt brought on by outmoded political, economic, and technological systems.

Decorum

Polite behavior has much to recommend it. However, unstinting politeness allows the system to be undermined by bad actors, who can use the system's very own protocols (civil servants, the courts) against it. When it comes to climate action, some friction may be necessary to achieve what Martin Luther King Jr. called "a positive peace which is the presence of justice."[15]

Distrust of Science

Some skepticism of science arises from legitimate concerns about limitations of certain research methodologies or the use of science in weaponry and pesticides. The real danger is distrust of the scientific method itself, and politically motivated resistance to science-based policies that have clear societal benefits in mitigating the climate crisis.[16] Ethics in science is essential to correcting misinformation.

Ideology

You can sidestep ideological divides by seeking diverse collaborators. But be wary of leaders who use ideology to divide and conquer. We can't wage an effective war on climate change if we are too busy fighting with each other.

Nature Blindness and Speciesism

Lack of familiarity with the natural environment and other-than-human species creates psychological distance that allows eco-destruction to proceed and worsen. Adopting any kind of naturalist practice, like birding, gardening, plant identification, traditional hunting and gathering activities, eco-arts, or nature photography, can help you to see beyond our humancentric world and tap into our interconnectedness with other beings.

NIMBYism

It is easier to advocate for environmental justice when it comes with no personal cost. For example, one might support policies and funding to relocate people who have lost their homes to climate disasters but feel resistant to hosting refugees in one's own community—what's referred to as a NIMBY ("not in my backyard") attitude. NIMBYism can be motivated by distrust, fear of others, entitlement, attachment to the status quo, and an unwillingness to sacrifice comfort. We need to check these tendencies at the door.

The Normative Bubble

We're more likely to engage in climate action when we believe that other people are already taking that action, yet most people drastically underestimate the level of public support for climate action worldwide. This gap is a "normative bubble" that must be popped.[17] Continually remind yourself and others that studies typically reveal support for climate action in the 80 percent range.

Poor Messaging

When social psychologist Robert Cialdini studied efforts to prevent visitors to Arizona's Petrified Forest National Park from stealing pieces of ancient fossilized wood for souvenirs, he found that signs admonishing park visitors about stealing ("Many past visitors have removed the petrified wood from the park, changing the state of the Petrified Forest") actually *increased* the undesirable behavior.[18] Why? Emphasizing the frequency of the fossil theft actually normalized it, and as we have seen, people tend to behave in accordance with social norms. Signs that reminded people that theft was *not* normal ("The vast majority of past visitors have left the petrified wood in the park, preserving the natural state of the Petrified Forest") were more effective. Signs that clearly identified what *should not be done* ("Please don't remove the petrified wood from the park") were most effective.

Reductionism

Viewing wicked problems like climate change through a single lens (such as evolution, economics, psychological biases, good-versus-evil narratives) is a form of reductionism that limits your ability to problem-solve in a holistic way. It's frequently said that "for every complex problem there is a simple answer that is wrong."[19] Think big picture.

Sunk Costs

The "sunk cost fallacy" is the tendency to continue a course of action you have invested time, money, effort, or emotional commitment in, even when abandoning it would be better.[20] If you are struggling to sustain motivation in your climate efforts, it's okay to abandon your current approach and redirect your time, energy, and passion toward a new goal.

The Peter Principle

This refers to a situation in which a qualified person continues to be given increasingly difficult tasks and responsibilities until they reach a level of incompetence. In one sense, eco and climate issues force all of us to take on problems we are not prepared or qualified for. A particular problem is when people imbued with power, wealth, or fame are placed in leadership roles that they are unqualified for. Humility, having a growth mindset, and accepting social support can help mitigate the harm this can cause.

Pseudoscience

Using the trappings of science to dress up opinions and beliefs that are unsupported by evidence is no less a barrier to action than the wholesale rejection of science and its methods. At best, the use of pseudoscience involves cherry-picking information to support your position and disregarding information that might prove you wrong. At worst, pseudo-scientific ideas are used in disinformation campaigns designed to discredit legitimate science. Learn, fact-check your sources, and do your homework. Don't be the one responsible for spreading misinformation.

Rationalism Bias

We tend to believe that everything can be explained and that people are always rational actors. This expectation produces the assumption that giving people accurate information about threats and solutions will lead to new behaviors and social change. Ironically, always expecting *rational behavior* is an irrational belief.

THE CLIMATE SURVIVOR: PERSONAL HEALTH AND INTEGRITY

This role is about you, your personal sustainability, and the private actions you take in your life. Here the goals focus on your own health, safety, and growth. The barriers are personal: impediments to your ability to cope, lack of connection with the land and a home place, patterns that keep you isolated, and beliefs and stories you tell that prevent you from feeling happy even when it might be appropriate.

Comfort

The desire to stay in your comfort zone can stem from the fear of discomfort or the fear of losing control over your comfort. In therapy lingo, it can be easy to get stuck in a "discomfort zone" where you're neither happy nor uncomfortable enough to motivate change.

The Dunning-Kruger Effect

This cognitive bias refers to the tendency to overestimate our abilities relative to others.[21] Variations include the "discovery problem" (believing you have discovered an idea or innovation when it is only new to you) and "planting a flag" (claiming a space—literal or figurative—as yours, being unaware of the previous or existing inhabitants). Remember that truly new ideas are rare, and you are not the first person to encounter the world.

Constraints

Even in the presence of motivation and intention, the two most common constraints on your eco-behaviors are the lack of options and the lack of behavior-specific knowledge. For example, the solar panels with which you want to power your home could be unaffordable (lack of options), or you could have access to solar panels but not know how to install them (lack of knowledge).

The Curse of Knowledge

When you are knowledgeable and passionate about environmental issues and are used to being in a context where others are as well, you can lose perspective.[22] When trying to educate newcomers about climate issues, don't assume they know everything you do, and be aware that the complexity, specialized jargon, and nuance that come with your expertise can be overwhelming and off-putting. Start simple. As the saying goes, if someone is thirsty, give them a drink of water—don't shoot them with a fire hose.

Déformation Professionnelle

The French term *déformation professionnelle* refers to the tendency to narrowly see and experience the world from the perspective of your job or profession.[23] When you bring together individuals from a wide variety of disciplines, fields, and crafts, each with their own ways of diagnosing the problem and their own approach to solutions, they may struggle to find common ground. To get over this barrier, see if you can widen your aperture and see the climate elephant from a new perspective.

Disgust and Moral Superiority

If you perceive Earth as sacred, you may feel judgment or even disgust toward those who understand Earth differently from you.[24] Don't let these feelings stop you from collaborating and finding common cause with people working toward positive outcomes.

Ego

Many environmentalists are forced to work in isolation and against great odds. Solitary struggles that require fierce determination and self-belief can also make for ego-focused views about the right action, and a sense that others are either "with me or against me." This also makes it hard to collaborate and recognize multiple views of the climate elephant. Try not to go at it alone.

False Summits

Mountain climbers are familiar with the experience of reaching a peak and discovering that the true summit is still farther away. This is a danger with wicked problems like climate change: Each time you make progress, you find there is more work to do. Keep at it. The more mountains you climb, the clearer your objectives become.

Fixed Mindset

As we've discussed, the belief that your skills, abilities, and options are fixed and can't be improved through effort and learning can contribute to feelings of inadequacy and despair. Remind yourself of how you have changed in the past and give yourself permission to do so again.

Hand-Waving

Don't wave away hard questions or valid criticisms about your plan for action. It can be tempting to gloss over gaps or inconsistencies, but a far better approach is to admit you don't have all the evidence or details yet, say that you are working on it, and then follow through.[25]

Illness

Be it a cold or cancer, when physical illness stops you in your tracks, you need to focus on your health. If severe emotional distress impedes your ability to make a difference, seek help. Honor the need for self-care in yourself and others.

Impatience

It takes time for governments and infrastructure to catch up with technologies and society's needs. The process of climate mitigation and adaptation is one of many stages in the development of human societies. It took hundreds of years for the city of London to put sewer systems in place to eliminate cholera epidemics. The pace of change is quicker now, but you can't expect to see progress overnight.

Giving Up

Until you reach a certain point, your efforts toward reaching a new goal may appear to be useless or ineffective. A lesson from whitewater rafting is "don't stop rowing." Sometimes it's the final stroke that gets you past the obstacle in your path.

Learned Helplessness

After a person has repeatedly experienced a stressful situation that they perceive themselves unable to control or change, they tend not to try even when opportunities for change are available. Remind yourself that even if you feel helpless in one situation, that doesn't mean you can't have agency in another.

The Moral Fallacy

Just because you consider an action to be moral or undertaken for good reasons, that doesn't necessarily mean the action is right or appropriate. Outcomes matter as much as intentions, and sometimes well-intentioned actions have unexpected consequences. Another variation of the moral fallacy is the assumption that because we behave ethically, others will too. Yet this blanket assumption is what allows bad actors to move in.

Overdoing It

Overwhelm is a normal emotional response reminding you that you can't do it all. If you try, you'll end up feeling like a jack of all trades and a master of none. Rather, be aware of a range of issues and actions and choose one area to get better at. Strive to be a jack of all trades and master of some.

Perfectionism

Holding yourself to unreasonably high standards will prevent you from pursuing good-enough projects that would make a difference for people and the planet. Practice eco-pragmatism instead of eco-perfectionism.

Shame

Imagine a mother who is excited to spend time with her children building a crossing of flat stones across a wooded creek, as she did when she was young—until a passing naturalist informs her that moving and reconfiguring river stones can destroy the sensitive ecosystems of salamanders and caddisflies and alter the creek flow, creating erosion. Suddenly the mother's earnest attempt at instilling appreciation of nature is transformed into a source of shame. As you continue to learn how best to take care of nature, you will make mistakes. See these as growth opportunities and not a negative commentary on you.

Social Rejection

Standing up for the reality of climate breakdown in the face of denial can be difficult. And the costs of having a view that is at odds with the view held by members of your community can be high. In these situations we have a tendency not to rock the boat. Seek membership in groups whose beliefs about the importance of eco and climate action align with your own.

Stubbornness

Stubbornness is a powerful trait if you can use it wisely. But it can also be a barrier to effective action when it takes the form of hardheadedness, clinging to what you know, a fixed mindset, ego, or pride.

Wanting to Win

Everyone likes to win an argument. But the reality is that people will always frame things according to their values and perspectives to preserve their self-esteem.[26] There will be no definitive judgment and no cosmic referee to say who is right. So, "winning" is rarely an option.

GETTING OUT OF A HOLE

Now it's time to be honest. Which of these barriers felt familiar to you? All of us fall into these traps at one time or another. The important question is which ones you are most vulnerable to. Which scare you or rob you of energy and motivation? Being proactively aware of these responses will prepare you to meet barriers and setbacks—and when you fall into a hole you can get out a little easier. Keep in mind that hitting a barrier is a *positive* sign. It means you are *confronting limitations*, either your own or the system's. You are on the threshold of change, and the frontier of what's possible. The key is to keep a growth mindset.

Sometimes you can overpower barriers; other times you can neutralize them; sometimes you can go around them. The point is to see them coming, so that you can say, "I already know." The next time someone breathlessly shares some indignation they have about global problems, you can have compassion for their suffering and know in your heart that you're already at work living the question "What is my duty, and how can I best fulfill it?"

Chapter 19

Strategy

I am asking, What kinds of shared futures can you and I imagine and bring into the realm of the possible, despite a highly organized investment in business as usual?

—Min Hyoung Song[1]

Strategizing about action in the face of global issues is daunting (never mind actually taking it). I understand. Being hesitant about the right action *you* should take, or the right direction for *you* to go, is normal given the scale of the problem and the stakes. It also stems from how we tend to see environmental problems (as so big) and our ability to effect change (as so small).[2] Since the dawn of the modern environmental movement in the 1970s, humans have been tasked with a new responsibility unique in history, to nurture and protect the *whole Earth*, not just their local air and water. We think that the scale of our efforts must expand with the scale of the problem. But this is backward. The actions you might take as part of surviving climate anxiety can start with whole-Earth consciousness, but they need to move on to something more tangible, more local, and more manageable.

If you've ever climbed a tree, you know the feeling of finding a sturdy branch to grasp. The branch feels so natural in your hands because your hand evolved to grasp it. We did not evolve to hold a fragile planet in our metaphorical hands. So, the first strategy for action is to stop trying to change or save the world, because you can't. Start trying to change or save one small part of the world because you can.

You've already taken some powerful actions regarding the world you can create for yourself. You laid the foundation for a personal sustainability pyramid, traced the timeline of your environmental identity, plotted a map of your losses and grief, envisioned a local place to stand and protect, and surrendered yourself to being bulldozed with an uprooted forest. You felt for your place in relation to the front line of activism. You imagined being a climate detective and climate champion.

There are many excellent guides specific to understanding scientific and policy solutions for environmental problems, detailed prescriptions on how to draw down greenhouse gas levels in the atmosphere, and calls to action for social change. In this chapter I share recommendations, advice, research findings, opinions, and approaches I have gathered that might be valuable for you while you are engaging in positive eco and climate activities, whatever they may be.

As I promised earlier, I wouldn't tell you what to think or to feel, but I would give you some options about how to think and how to understand your feelings. The same holds for taking action. In this chapter I offer a menu of options meant to help you look from new angles and refresh your creativity and perseverance.

We'll start with an exercise that grants you grace and forgiveness.

THE ECO-CONFESSIONAL

The eco-confessional is a ritual where people share personal stories of ecological irresponsibility. I learned of it from environmental educators Mitch and Cindy Thomashow. The exercise is meant not as a punishment or shaming but rather as a way to unburden yourself from the weight of the environmental transgressions you carry. It's an effective way to let go of perfectionism and appreciate that you do not bear responsibility for the challenges Earth is experiencing. And that process begins with a confession.

In his book *Ecological Identity*, Mitch tells the story of a time a few decades ago when he, Cindy, and their young children spent the summer on a remote island off the coast of Maine. Facilities on the island were rustic, without any trash or waste removal service, and they faced a dilemma of what to do with the dirty diapers that had accumulated by the end of their trip. So they copied the method used by the locals and threw their plastic bags full of diapers into the sea to be taken out by the tide. However, they misjudged the currents and the bags lingered near shore. Soon seagulls pecked open the bags, scattering the diapers around the harbor, where they floated "like white buoys…marking a trail of neglect." Mitch confessed that he was "profoundly embarrassed, suffering enormous shame, feeling like a first-class jerk, and a moral hypocrite."[3]

I facilitated this exercise with Liberty and Sam's Parents for Climate group and also coached Cassandra to use it in her teaching and ministry. As they often do, the collective discussions inevitably moved from sheepish laughter, to expressions of sadness and shame, and ultimately, to catharsis and relief. People shared the many ways they betrayed their own eco-values: throwing away things that could be recycled, driving to work instead of taking public transportation, taking carbon-polluting flights, even pursuing careers in industries or companies that perpetrated harm on the environment. Even in family and intimate relationship settings, the eco-confessional can help clear the air, as it did for Sally and Martin's recycling debates.

In a community of climate hostages, nearly everyone harbors something they feel uneasy about, whether it's our daily acts of wastefulness or our deeper regrets over what more we *could* have done to help protect the planet. Society doesn't provide a process to forgive our eco-sins. If anything, it seems to encourage them. So we need to forgive ourselves. When you do, you unlock a whole swath of healthy emotions—apology, compassion, forgiveness, repentance, clemency, and grace. In the Parents for Climate group, the energy of releasing the pent-up feelings was palpable, as was the desire to make amends and the excitement to try again.

ACTIONS TO TAKE

Unburdening your conscience frees up mental energy for taking action. But what then? One of the most common barriers to environmental action is a lack of clarity about which actions can have true impact. Here's a simple but inclusive framework from which to view environmental behaviors that helps keep you honest about the true impact of your efforts and sets you up for real, substantive change.

Ceremonial Behaviors

Christina knew that kneeling and crossing herself was a way to show respect at the sacred sites she visited during her pilgrimage on the Via di Francesco, but these gestures alone did not make her a good Catholic. Similarly, Sally owned several eco-friendly reusable grocery bags, but that alone did not make him a good environmental steward. *Ceremonial behaviors* are acts that have limited impact and are done to express your values and integrity. They could range from picking up litter to decorating your living space to express your environmental identity and signaling your policy positions through T-shirt and bumper sticker slogans.

Don't dismiss these small symbolic acts. They have an impact because they reinforce your eco-identity and bolster your self-esteem and your sense of agency when other actions have little impact. The key is to be aware that these behaviors are symbolic, and make them just one part of a larger suite of actions and efforts, including activities that require more patience, effort, and sacrifice.

Low-Hanging Fruit

Some pro-environmental actions are the equivalent of *low-hanging fruit*: they are simple, relatively low-cost, and easy to reach for. On a personal level, this might include shopping for secondhand clothing instead of fast fashion, avoiding plastic containers, and, if you have the means, purchasing an electric vehicle or a heat pump. For organizations, it can include bringing recycling and sustainability to the workplace, offering incentives to work from home and use public transportation, and other efforts to reduce energy use. Low-hanging-fruit behaviors have strategic value as the logical first steps toward more powerful changes. But don't become dependent on them, as they can be convenient substitutes for taking on more complicated and difficult challenges. As a society, we have been leaning on these low-hanging-fruit behaviors for far too long.

Real Change

I use the catchphrase "real change" for substantive efforts that will ultimately scale up to address the main drivers of climate change, such as greenhouse gas emissions, deforestation, industrial livestock farming, et cetera. Generally, these efforts require some personal sacrifice. For individuals and families, that can include adopting a plant-based diet, retiring appliances fueled by methane (natural gas) and vehicles that use fossil fuels, transitioning to solar or wind energy sources, and reducing nonessential air travel or offsetting the emissions.[4] For businesses, organizations, and governments, this entails honestly recognizing *externalities*, the environmental harms that aren't factored into the cost of goods and services, and accounting for the full scope of carbon emissions and other harms across the supply chain, energy use, and life cycle of products (also known as scope 1, 2, and 3 emissions).[5]

At the societal level, real action includes legal efforts to compel enforcement of environmental laws on the part of governments and corporations and to protect and ensure citizens' rights to a safe and healthy environment. There are organizations, like the Citizens' Climate Lobby, which seek to influence legislators to hasten the creation of carbon markets and emissions trading. International efforts include sustainable economic development, reparations for climate damages, and sharing improved technologies for climate mitigation and adaptation.[6] You don't need to be a lawyer or lobbyist to play a role in this kind of change, Anyone can contribute by supporting these organizations and causes.

In the absence of a needed social and government response, real action can also take the form of nonviolent civil protest to draw attention to the issues and direct action to block and disable new fossil fuel infrastructure. With sufficient international will to enforce a global transition to a safer and sustainable low-carbon economy, we could imagine an international program of eco-defense, with nations and coalitions employing economic sanctions to enforce environmental laws, implement blockades of fossil fuel exports, and dismantle fossil fuel infrastructure. While this kind of unified global effort could be seen as wishful thinking, it is quite realistic. Similar international enforcement mechanisms are in place to limit proliferation of nuclear weapons, prevent the illegal production and trade of chlorofluoro-carbons (CFCs) and other chemicals banned by international law, prevent human trafficking, maintain a moratorium on commercial whaling, stop illegal wildlife trade, and uphold other international agreements.[7] Here too you can play a role, by voting for representatives of government who support environmental regulation and enforcement.

ACTIONS TO CALL OUT AND RESIST

Unfortunately, efforts at real change often take a detour toward *climate theater* and *bad-faith action*, both of which are situations we must try to avoid.

Climate Theater

Public slogans, declarations, and advertising that suggest systemic, large-scale change while remaining ceremonial, ineffective, or nonexistent constitute forms of *climate theater*. These performative actions let companies and governments off the hook and are often deployed when being truthful about climate impacts threatens profitability or the organization's image or brand. Climate theater includes commitments to reduce emissions without follow-through, *greenwashing* (marketing products or initiatives as more earth-friendly or sustainable than they are), or even hosting climate meetings that require carbon-intensive travel.

On an individual level, climate theater could mean making a public show of practicing sustainability (like charitable giving or driving an electric vehicle) while simultaneously ignoring or minimizing the negative impacts you are having in other areas (like heavy emissions from frequent air travel).

Climate theater and other performative actions may look good and even assuage guilt, but they are deceptive and provide a cover for continuing business as usual. Be willing to call out climate theater when you see it.

Bad-Faith Action

Unfortunately, many people, companies, and governments are complicit in intentional, bad-faith actions to prevent or sabotage climate action.

This includes discrediting scientific findings about the clear and present dangers of unchecked fossil fuel use, false advertising of harmful products (marketing climate- and health-destructive methane gas as "natural gas"), and violating environmental regulations (as in the extraordinary case of the Volkswagen corporation equipping over half a million cars with devices that allowed them to cheat US auto emission standards).[8] This is the malfeasance that leaves you in the climate hostage situation, and it needs to be exposed and countered using legal action, education, and other strategies.

The persistence of premeditated bad-faith action, like the persistence of climate theater, leaves us all exposed to increasingly deadly and costly climate disasters (unnatural weather, drought, wildfire, flooding, sea level rise, disease, invasive species, displaced peoples, regional conflicts, etc.), which in turn breeds anxiety, misinformed helplessness, *doomism*, and reactionary politics. This is what the progression from ceremonial actions to low-hanging fruit to real change will prevent.

FINDING THE RIGHT SCALE FOR YOUR ACTIONS

When working on solutions, remember that you don't need to solve *everything* about climate change to solve some things. In fact, "solving" is not even the most appropriate goal for most wicked problems. Rather, the objectives include engaging mindfully with the issues, reducing harm, and creating local and system-wide benefits as you can.

When I am working with young people, I have them make two right angles with their thumbs and forefingers and bring them together to create a window that they can look at the world through. I tell them this is their viewfinder to help them select specific issues, species, or places to focus on. Once they have zoomed in on an objective, I tell them, they can zoom out to see what that action might look like at larger scales.

I encourage you to be curious about the scale of what you are trying to achieve, and to not limit yourself. The "we are all to blame" myth of climate propaganda has done a terribly good job keeping people focused on ceremonial actions, low-hanging fruit, and climate theater. But real change is not out of reach.

Start with Physical Scale

There is a rule I have for judging the potential effectiveness of actions: If an action can't be effective *somewhere*, it's unlikely it will be effective *anywhere*. So a question is, what specific place can your action or project take root in and grow from (beyond your imagination)? Slade made a point of searching for successful examples of solutions for water quality and habitat protections that might be adapted to the lands he was stewarding.[9] He started with small prototypes developed in cooperation with local Native tribes and community members, and expanded from there. Beware of *nowhere ideas*— abstract projects or goals that are divorced from the land, and from real places and living ecosystems. Make your actions *somewhere ideas*.

Then Consider Time Scale

Improving your personal health, fitness, and work performance or strengthening intimate relationships requires sustained habits of action over a long period of time. To scale up your sustainability efforts, it's helpful to start with short practices that take a couple of minutes. Marcus briefly went outside each morning before looking at news or screens, paused to breathe and clear his head after engaging in challenging work, and asked himself what he would like to feel as a way to sustain his energy and motivation.

To make progress on tasks that they struggled to fit into their busy schedule, such as writing letters to their legislators or organizing an event or fundraiser, Liberty and Sam applied the "twenty-minute rule." They set a timer and committed to twenty minutes of effort. This made the best use of their time together and often allowed them to get into a mindset for a longer, focused work session.

To increase the scale of your learning or your impacts, more time is needed. But not as much as you might think. The "twenty-hour rule" is another good approach, since typically twenty hours of focused practice is required to gain basic competency with a new skill or activity. Christina used this approach to gain hiking and outdoor skills, educate herself about the routes for her pilgrimage trek, and practice her Italian. You can reserve the classic "10,000-hour rule" for environmental vocations you want to master and that you're willing to devote some years of your life to, like Delphina earning her veterinary degree. Other examples include building a business or nonprofit focused on eco-issues, running for office, or establishing an organic farm. Programs like 80,000 Hours (which takes its name from the length of a typical career) use effective altruism principles to help people leverage their time and efforts at an even larger time scale by finding a career that solves global problems with high need and fewer resources devoted to them. Jann consulted the 80,000 Hours career library when planning her next chapter.[10]

Next, Prioritize

In his book *Four Thousand Weeks*, Oliver Burkeman describes a number of time management principles that can be applied to your sustainability or climate action. For example, "serialize projects" by selecting one life duty at a time and see it to completion before moving on to the next, as Roman and Sandra did when they agreed to set major life decisions aside until Roman completed his dissertation on seawater intrusion under the Thwaites Glacier. To make progress on your eco-goals, you can also practice what Burkeman calls *strategic underachievement* and identify areas of your life in which you won't expect excellence from yourself, so you can focus on areas that are most important to you or where you can have the most impact. While Sally and Martin typically devoted time to landscaping their home and garden, they consciously let their yard get overgrown while they focused on higher-priority goals, like retrofitting their house with electric appliances or doing their solar power and plastics reduction advocacy work.

Practice "Goal Set-Through"

Make sure to set your eco or climate action with an eye toward future, ongoing goals, so that the energy invested in one objective flows through to the next. For example, the goal a tennis player should set is not just to win a set but to win the match, and not just to win the match but also the tournament, and to remain healthy for future competitions. Goal set-through was helpful for James to get over his tendency to put all his hopes into singular projects and campaigns and then find himself adrift when they finished or if it didn't work out.

Add the People and Relationship Scale

A general approach to effective action is to "be an agent of change in your circle of influence."[11] That is, start where you are, with your existing relationships, and go from there. This includes family, friends, neighbors, and coworkers, and can potentially extend quite far depending on the size of your network. But resist the need to be an "influencer" competing for followers. A short-term strategy is to focus on the *quality* of the relationships you cultivate, not the quantity. Rick found this a satisfying way to approach mentoring young people, instead of his traditional data- and results-focused approach.

Sometimes, to make progress on your eco and climate actions, it's not *what* you know but *whom* you know. The greatest idea is not worth much if you are alone and cannot share it. Even small ideas and actions can have large effects when combined with the actions of others. One of the rewards of taking action is the communal synergy you feel when you meet others who spark your imagination and allow you to take on climate issues at larger and larger scales.

Stick Together

As sustainability scientist Hannah Ritchie has illustrated, it is important to stick with those who are working in the same direction regarding a safe and healthy environment, even if they favor different methods. Remember that even those "pushing at a slightly different angle" are on the same team, as Ritchie puts it.[12] Perfectionism, ego, and infighting dissipate collective energy and allow those who oppose environmental action to gain ground. Keep inevitable debates generous and productive. Look beyond your particular group or belief system to see the larger picture. Solutions don't need to be your way or the highway. Join people who are working together across different approaches to achieve the common good.

Stick with Others Pushing in the Same Direction

"Those pushing in the same direction, even if at a slightly different angle, are on the same team." - Hannah Ritchie

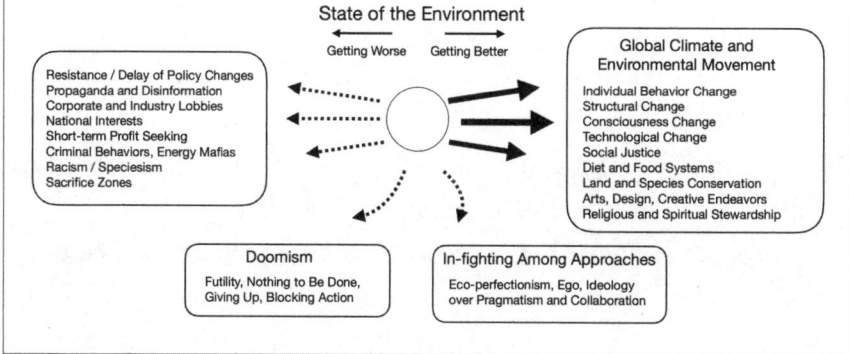

State of the Environment

← →
Getting Worse   Getting Better

Resistance / Delay of Policy Changes
Propaganda and Disinformation
Corporate and Industry Lobbies
National Interests
Short-term Profit Seeking
Criminal Behaviors, Energy Mafias
Racism / Speciesism
Sacrifice Zones

Global Climate and Environmental Movement

Individual Behavior Change
Structural Change
Consciousness Change
Technological Change
Social Justice
Diet and Food Systems
Land and Species Conservation
Arts, Design, Creative Endeavors
Religious and Spiritual Stewardship

Doomism

Futility, Nothing to Be Done, Giving Up, Blocking Action

In-fighting Among Approaches

Eco-perfectionism, Ego, Ideology over Pragmatism and Collaboration

Have an Exit Strategy

In addition to committing *to* projects, commit to *let go* of projects that are not working or unhealthy for you. Be honest and change course if the evidence calls for it. Adaptability is as important as the ability to persist. It's okay to bet on a promising new approach, and lessons from bold attempts that falter can often teach you more than safe wins.[13] Ethical ways to create exit strategies include enlisting in an action for a set amount of time or signing on for a discrete project. Of course, not everyone has choices about what they need to deal with or how long they will need to deal with it. Jesse couldn't postpone working through the aftermath of the wildfire on her family land. Liberty couldn't quit in the middle of her cancer treatment. In these situations, rather than feeling condemned to action, maintain a growth mindset. Let go of old habits and try new strategies. As we discussed earlier, try to have agency within your agency.

TOOLS AND STRATEGIES

No matter what climate issues you choose to work on, there are tools and strategies you can adopt to conserve your energy and amplify your impact.

Using Technology

How can you use technology to help meet your eco and climate goals? One way is by taking an activity or process and systematizing and automating it so that it becomes faster or more powerful (for example, you can use an app to track your carbon footprint, or software to manage your organization's fundraising efforts). An astute observation I've seen about the pros and cons of technology is attributed to Bill Gates, and is paraphrased as: Technology will make an efficient process more efficient, and it will make an inefficient one more inefficient.[14] It all depends on how you use it.

When Larysa enlisted in her local Community Emergency Response Team, she found that the best way to develop her skills wasn't online or via some learning software but through hands-on work with her neighbors and local officials. Much like with cooking, art, gardening, or any other skill, work with your own hands first, and then add technology.

Design Thinking

Problem-solving approaches drawn from the arts can be helpful for envisioning solutions to whatever eco-issue you decide to tackle. A design-thinking approach involves focusing less on the details of problems and more on what solutions would look like, particularly when put into practice within people's daily lives and experiences. This is a process-oriented approach that starts with observing and questioning and then moves on to brainstorming options, creating prototypes, and continually testing and improving them.[15] Solutions aren't imposed from the outside of a system with the hope they'll work. They are built step by step, gaining feedback and iterating until you discover what works.

We began the book with the recognition that when you're thinking big, you're likely not thinking big enough. As we've seen, eco and climate issues often manifest as wicked problems, multiple entwined issues or dilemmas that resist any straightforward fix. Design thinking opens us up to the possibility of meeting these complex problems with wicked solutions. We don't fix; we respond and engage with the situation, working within the constraints and testing ways to make parts of the problem better without making others worse. A wicked solution might not look obvious or elegant, but it works well enough to move you forward to the next phase of problem-solving. Pragmatic climate legislation like the 2022 Inflation Reduction Act (IRA) is a great example of a wicked solution. While far from the full-scale economic and societal mobilization that many hoped for, the IRA made advancements on wicked climate problems as varied as greenhouse gas emissions, electric vehicle and solar energy technology, infrastructure investments, jobs, sustainable agriculture, and local climate adaptation.

Oblique Strategies

Oblique Strategies is a collection of thought-provoking prompts, metaphors, and suggestions that originally was devised to help artists and musicians overcome creative blocks but which are now used more widely to encourage creative, out-of-the-box thinking.[16] "Oblique" suggests viewing or approaching whatever you are working on from the side, from a new angle, or with a fresh perspective. Examples of Oblique Strategies adapted for climate coping or taking on wicked environmental problems could include:

- Define an area as "safe" and use it as an anchor. (Find a part of your project that is solid, and work from there; a local place where you can restore yourself; or a coping skill that helps you feel safe in your body.)
- Go slowly all the way around the outside. (Take your time. Think widely about a new challenge or an endeavor to get a perspective.)
- Get many opinions. (Be open to learning other ways people look at the climate elephant.)
- Determine what scale you are operating from. (Personal life, family, community, wider culture? Your own backyard, literally, or a region, or beyond?)
- Ask your body what it needs. (Rest, nourishment, care, movement…?)
- Go outdoors and ask the same questions again. (How do things look and feel with your feet on the ground and the sky overhead?

There are things you just can't see from indoors—either beauty
you're missing or issues you're sheltered from.)

- What would you advise a best friend in the same situation?
  (Direct your compassion and wisdom to yourself.)

Storytelling

You can't coordinate effective action without effective communication and
storytelling. And no matter what story you are telling, one of the most
important things is to find a style that is genuine for you and comes from
your life. Because I am a psychologist and therapist, my style is supportive and straightforward, expresses care, and also is as direct and truthful
as possible. When I was studying literature years ago, one of my teachers,
Kenneth Koch, who was a figure in the New York School of poetry, recommended that writers "be serious but not solemn." That advice is something I've carried with me. Writing about environmental degradation and
climate breakdown naturally lends itself to solemnity and melancholy and
can easily become pious and judgmental. Resist the tendency to become
self-important, gloomy, and sanctimonious, as this can lead to "preaching
to the choir" (wasting your words on people who already share your beliefs)
or turning people off.

Marine biologist turned filmmaker Randy Olson stresses the importance of having a clear narrative and the building blocks of a good story.[17]
People's brains show similar activity, or become aligned, when they hear the
same story. Olson identifies four ways to create such alignment: by connecting with your audience's "head, heart, guts and groin." The first is easy: It's
tapping into their intelligence and rationality through the sharing of information. The heart, of course, is feelings, such as sympathy and love. The gut
is further from our head; it's where we tap into our intuition and instincts.
And the groin is where we feel passion and excitement. If your message is
sexy enough, it can connect with most anyone.

Not surprisingly, science and climate communication generally comes
from the head and tends to dismiss or ignore the importance of expressing and validating feelings, appealing to people's better instincts, or igniting

their passion. But this need not be so. You can simultaneously share information, communicate your gut level values, *and* transmit your passion and energy when you tell your eco-story.

LEGACY AND LIFE-TRANSCENDENT GOALS

Don't shy away from pursuing ambitious goals that are larger than your life. You never know how far your efforts might reach. Here's a quick thought exercise to help you understand the power of legacy: Imagine that you were an abolitionist, someone who wanted to put an end to slavery, born in the United States around 1780. You lived a long life for that era, dying sometime around 1860, on the eve of the Civil War and several years before the Emancipation Proclamation. At the end of your life, you might have considered yourself a failure. But, from our perspective today, was this person's work for naught? The truth is, we don't know what will happen after our time. We can't always see the full scope of our contributions on a larger historical scale.

Remind yourself that the actions you take are continuing the legacy of others who came before you and will become your legacy for future people to carry on. If you live your life in the context of seven generations, that is a span of approximately 175 years. You'll do what you can in your life, and others will take up the work. This is a powerful way of coping during dark times when progress is undone. Resist judging your efforts by today's outcomes. Instead, practice committing to causes and programs that will outlast you. Then, all you must do is show up and extend your eco-timeline, one day at a time.

Chapter 20

Weathering a Disaster

Wasn't that a mighty storm
That blew all the people all away.

—TRADITIONAL FOLK SONG

—Traditional folk song[1]

IN THE FACE OF CLIMATE AND ENVIRONMENTAL DISASTERS, all the dangers implicit in climate anxiety go from abstract ideas to present reality. The quote often attributed to William Gibson about the spread of technological innovations, "The future is already here—it's just not evenly distributed," is also a fitting description of the calamities associated with climate breakdown. In fact, we are all in the climate disaster cycle right now: either amid a disaster, recovering from one, or anticipating whichever one is to come next.

The harsh reality is that climate-change-fueled disasters are a fact of life now. Larysa and her family faced storms, wildfire smoke, and heat waves that made them fear for the health of their newborn. Jesse and her family were thrust onto the front lines of global heating by a supercharged wildfire that destroyed homes, property, and wild lands in Oregon. And when Reid and Marisol's families on the Gulf Coast of Texas were inundated by unprecedented storm surges and flooding, they needed to quickly evacuate and work together with members of their community to support others who needed help. When disasters arrive on your doorstep, what coping actions are needed?

We round out Part Five, on action, with a primer on skills needed to cope mentally and emotionally during an actual disaster. During the course of creating this book, disaster mental health went from an optional, value-added topic to a primary focus. It may even be your entry point. Along with strategies to help you prepare for a disaster, and tips on how to cope in real time, we will also consider the possibilities for new beginnings that disasters afford.

TIPPING POINTS

In 1912, when the ocean liner *Titanic* set sail on its first and only voyage, the ship was believed to be unsinkable. Its designers claimed that the sixteen separate sections in the hull ensured that if one area of the ship was breached and flooded, the other sections would remain buoyant. The ship was even designed to stay safely afloat even if multiple sections were flooded—for example, both the bow and the stern, or two or three compartments in the middle of the ship. These innovations, along with the new Marconi telegraph that could communicate with potential rescue ships, meant that a damaged *Titanic* could serve as its own lifeboat.

Unfortunately, when the ship struck an Atlantic iceberg in calm seas late on the night of April 14 of that year, the underwater mass of the berg sliced obliquely through six compartments, setting into motion the events that led to the *Titanic*'s rapid (two hours and forty minutes) sinking. As the forward compartments filled with seawater, the bow of the great ship sank deeper into the ocean, which then allowed water to pour into the aft compartments in succession, with the lighter stern rising high into the clear night air.

There is a scene in the Hollywood dramatization of the *Titanic* disaster when the ship's Scottish designer learns of the extent of the damage and declares to the incredulous captain that it is a "mathematical certainty" that the ship will sink.[2] These questions of mathematical certainty are what keep climate scientists awake at night. Earth has various planet-wide natural systems and cycles: carbon, oxygen, ocean currents, and so on. When the long-standing natural boundaries that govern these cycles are breached (as when ocean temperatures rapidly rise) we face unpredictable climate tipping points and feedback loops. While it is impossible to predict exactly what will occur, we can know with mathematical certainty the next phases will be rapid and likely destructive and deadly.[3] This is why you will hear terms like "uncharted territory" and "unnatural disasters." We are in a situation no one has ever witnessed firsthand in the history of humanity, and it is novel even in the history of the planet.

To put it mildly, coping with all this uncertainty requires a certain mindset.

A GROWTH MINDSET FOR DISASTERS

The great rock-and-roll drummer Neil Peart had a routine in his later years of touring. Far in advance of a tour, he would train physically to build his fitness and capacity, then he would rehearse the songs and drum flourishes by himself. Only after he had once again mastered the fundamentals would he start rehearsing with his bandmates, who joked that he needed to rehearse to rehearse. There is an analogy to weathering disasters here. It helps to build a healthy mental and physical foundation and to rehearse strategies for coping, to ensure you have mastered the fundamentals of preparedness when disasters strike.

Remember that a fixed mindset is when you consider your abilities, intelligence, and talents to be limited. A growth mindset, in contrast, holds that your abilities, intelligence, and talents are changeable. You can improve and add to them through your own efforts. Disasters are out of the ordinary events that by definition are overwhelming, and a state of overwhelm can create a fixed mindset response. The more prepared you are, the less overwhelmed you will feel. The less overwhelmed you are, the better your ability to grow and adapt. To be forewarned is to be forearmed. While extreme events all arise from unique circumstances, there are many aspects of disasters we can anticipate in advance.

ADAPTIVE ECO-ANXIETY

We've discussed how the eco-anxiety you may feel can range from a normal and healthy apprehension of environmental threats to a debilitating situation in which your anxiety becomes obsessive and begins to invade all aspects of your life. When I was able to support Larysa during an actual disaster—a relatively minor ice storm that caused a few days of road closures and loss of electrical power—we were able to experiment with a middle ideal of *adaptive eco-anxiety:* an anxious emotional response grounded in an accurate appraisal of the current situation and threats. This helped Larysa find a middle ground between the extreme anxiety that overwhelmed her and impeded her response (we called this "eco-panic") and, at the other end of the spectrum, an *insufficient* amount of anxiety for the situation that would undermine her ability to anticipate or react to true environmental threats. I called this situation *eco-naivete.* Larysa called this eco-cluelessness.

I find that many people are comforted by the idea of adaptive eco-anxiety. Once you understand that your anxiety is useful—at least to a degree—you can take a balanced view of your disaster risks and buffers, allowing you to calibrate your apprehension to the situation at hand.

The *Titanic* story is a great analogy for many aspects of climate change: the hubris of the designers, technologists, and investors; the discounting of the forces of nature; the voice of science trying to convince the leaders on the bridge to change the ship's course before it was too late; the rich enjoying opulent and possibly final meals; and the poor trapped below decks. But the lesson for living in an era of climate change is not that we are on the *Titanic.* We are on many *Titanics,* in many places, embedded in many systems, facing many perils. By this I mean that the nature, severity, and impact of climate disasters are hyperlocal; no two people will experience them in the exact same way. Tragically, many local *Titanics* will go down, often as quickly and catastrophically as their namesake. But the upside of this view is that there are many ships you can save. And at any given time, you only have to worry about the iceberg you are facing.

Ask for Help

If you are overwhelmed by taking on disaster scenarios on your own or if this is causing you significant distress, talk to someone who also takes these seriously. You are not alone, and this can be reassuring. This is also a time when getting assistance from a mental health professional or a personal coach is helpful. They don't need to be an expert on climate science, but they can help you to better calibrate the proper focus and intensity for your anxiety.

ANTICIPATING THE PHASES OF DISASTERS

Environmental disasters have distinct stages and often follow a similar sequence. They progress slowly, sometimes imperceptibly, until the crisis hits. Then things happen very quickly. There is often what is considered a *heroic phase*, where people go to extremes to save themselves and others, followed by a brief *post-disaster honeymoon* period of community togetherness as people work together to help each other and rescue survivors. Next is the *aftermath*—a period of disillusionment in which the extent of the damage is revealed. Finally comes *reconstruction*.

During the post-disaster period, there will be an inventory of losses and fatalities, and an investigation of the event, including retelling from different sides (one might say from different perspectives of the climate elephant): first responders, local community members, and commentators from the outside. For severe disasters, especially when there were lapses in human judgment that resulted in a community or region being caught off guard, there will be a reckoning—official investigations, trial by media, and informal finger-pointing in the local community—including fault finding and a demand for justice.[4]

When large and well-publicized disasters unfold in real time, multiple phases occur simultaneously. During the 2025 Los Angeles wildfires, even while heroic efforts to contain the blazes continued and community support flowed in, a reckoning commenced, seeking leadership during the crisis and to place blame and responsibility. Unfortunately, the ongoing forces of politicization and climate propaganda quickly sought to "spin" the disaster—in this case, away from the reasons the wildfires were so unnaturally severe: global heating, drought, and amplified desert winds. For some hard-hit places, the disaster phases can be protracted and chronic, a cycle of perpetual disaster and recovery, with the toll of multiple disaster events overlaid on top of each other, preventing a stage of community recovery and reconstruction.[5]

Phases of Emotional and Psychological Responses to Disasters

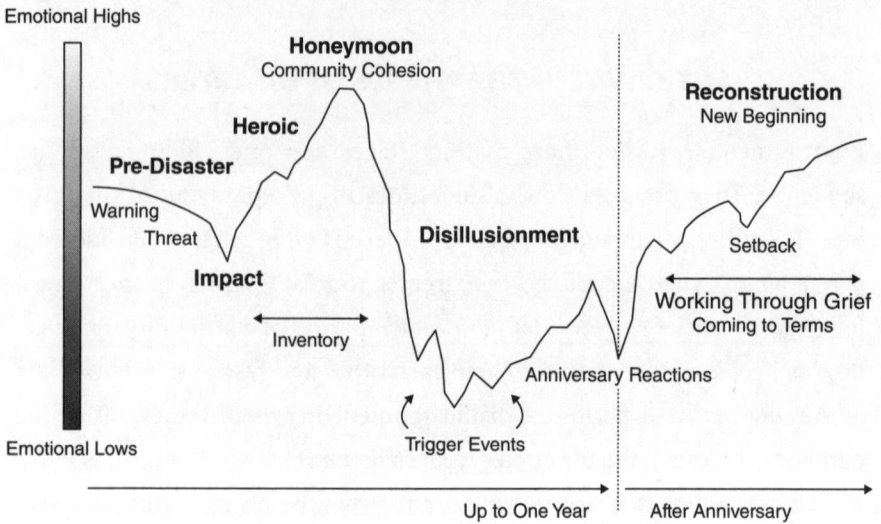

Emotional Highs

Honeymoon
Community Cohesion

Reconstruction
New Beginning

Heroic

Pre-Disaster

Warning

Threat

Disillusionment

Setback

Impact

Working Through Grief
Coming to Terms

Inventory

Anniversary Reactions

Emotional Lows

Trigger Events

Up to One Year

After Anniversary

AFTER THE DISASTER: A PARADISE MADE
IN HELL OR DISASTER CAPITALISM?

It is true, as we discussed earlier, that disruptive events can bring communities together, especially disasters that are perceived as "natural" and don't have a clear human cause or culprit. Rebecca Solnit highlighted this phenomenon in her 2010 book *A Paradise Made in Hell*. Disasters inspire moments of altruism, resourcefulness, and generosity. Amid all the loss and heartbreak, a disaster is an *unfreezing* of the system that can provide a sudden opening for new possibilities that have been waiting or which are invented in the moment. For a while there may be hopeful visions for a post-disaster life. Winston Churchill is reported to have said, during negotiations near the end of World War II that led to the founding of the United Nations: "Don't let a good crisis go to waste." While the disaster itself is not in your control, what you think and do about the disaster is.

Unfortunately, disasters also reveal faults and vulnerabilities in communities—such as social class divides and the legacy of past injustices— and these tensions are especially likely to flare up in instances where impacts, aids, and support are allocated unevenly. A community that is at war with itself over scarce resources is vulnerable to being exploited by opportunistic local players or outsiders. The phenomenon of *disaster capitalism* arises when the shock of the emergency and disorganization of government systems provide an opportunity to privatize locally managed public services or skirt the typical regulations and safeguards.[6] Racism and violence can be stoked as well. During the 2022 wildfires in Oregon, armed vigilantes set up roadblocks in some rural areas based on stories that radicals from the city had either started fires or were looting homes that had been evacuated.[7] While false, these rumors inflamed fears and created very real dangers.

The success of any disaster recovery hinges largely on how people channel the positive energy that comes from collective coping, and the degree to which they can prevent unhealthy responses like disinformation, violence, and profiteering.

CREATING A NEW BASELINE FOR UNNATURAL DISASTERS

You can't rely on memories of past disasters to guide you.

As children growing up on the Gulf Coast, Reid and his wife, Marisol, found that the run-up to a big tropical storm or a hurricane was both scary and exciting. The post-storm times were associated with check-ins with neighbors and community-wide rebuilding. Those who fared better helped those who didn't, and there was a deep sense of bonding. I have similar childhood memories of the great lake-effect snowstorms and blizzards that blanketed my home in Buffalo, New York, with roads closed to all but emergency vehicles and snow machines, walking across town with my father to his work, and witnessing the city digging out.

While these past disaster events are important points on your environmental identity timeline, be careful with nostalgia. Reid's family in the Houston area had many stories of Gulf storms. His great-grandfather had memories of the great storm of 1900, with its fifteen-foot storm surge that wiped away the low-lying boomtown of Galveston, Texas.[8] More recently, people referenced Hurricane Ike in 2008, a large Category 4 storm that took a track similar to the 1900 hurricane, and Hurricane Harvey in 2017, which dumped over fifty inches of rain on Houston over a four-day period (as much rain as the area usually receives in a year), yet neither storm produced a surge of more than thirteen feet at its peak.[9]

Using these past hurricanes as their baseline, Reid's family struggled to imagine scientists' storm surge projections of up to twenty-five feet. However, Marisol knew that the storm surge during Hurricane Katrina had been as much as twenty-seven feet. And Reid saw that the populated Houston metro area sits just north of Galveston Bay. The massive fossil fuel and chemicals infrastructure of the area—which accounts for a significant percentage of total US oil refining capacity—would be partially, if not totally, inundated by the modeled storm surge.[10] If this were to happen, it is estimated that over 90 million gallons of oil and hazardous substances would be released into the bay. So Reid and Marisol foresaw a combined Hurricane Katrina and Deepwater Horizon disaster as a real possibility.[11]

All of this led to the decision to relocate north with their children to Dallas, as we learned in Chapter 4. While Reid felt guilty about leaving the Gulf Coast he loved, he felt a sense of reassurance and safety that he was establishing an inland base to serve as a ready evacuation or relocation point for his extended family on the coast come the next storms.

THE EMPTY CHAIR: REMEMBERING THOSE MOST HARMED

In the United States, we are unconsciously buffered from the true toll of climate disasters by the disparate geographic locations that are impacted. We forget that before making landfall in Texas or Florida, tropical storms first touched down in places like Honduras, Turks and Caicos, and Cuba—often with deadly consequences. In the United States we also take for granted that we have access to certain resources, ones that more vulnerable nations lack, like the National Guard, the Coast Guard, the Federal Emergency Management Agency (FEMA), and thousands of local and national nonprofit organizations.

Reid and Marisol had volunteered with migrants in the colonias of south Texas and knew that while Houston faced dangers, places near the border were much more vulnerable to climate threats due to poverty and poor infrastructure. As part of their climate migration plan, they committed to donating to and volunteering their time with other, less privileged climate migrants through Marisol's Christian church group, and they planned to do a service trip to the Caribbean to help with disaster recovery.[12]

In times of climate disaster, taking action on behalf of the most harmed often has what I call instant karma, or what's called in environmental psychology "co-benefits": It helps you in the short term while also helping you and others in the long term. This is true in disaster response. As Reid and Marisol learned, when you care for others, it increases your sense of control, ability, and meaning, and helps you to recover more quickly.

PRACTICAL DISASTER PREP

A hallmark of a growth mindset is your ability to educate yourself. In Chapter 4, I led you through some ways to take control of your own local climate adaptation with your own "Individual Problems with Climate Change" report. One way to educate yourself is through books and resources like David Pogue's *How to Prepare for Climate Change*, Nick Mott and Justin Angle's *This Is Wildfire*, Haskell Moore's *Hurricane Preparedness for the Home and Family*, and guides from organizations like Fire Safe Councils and the Firewise Communities program. Of course, practical steps like assembling disaster "go bags" and strengthening your home against storms or wildfires are important components of disaster prevention. But these measures are just an entry into building a practice of disaster awareness that will need to *change over time* as climate breakdown progresses and our baseline of what constitutes a disaster changes.

BEING A COMMUNITY EMERGENCY RESPONDER

Often, helping others recover from disasters can be its own form of therapy.

Larysa experienced this firsthand after joining a Community Emergency Response Team through a Washington state program that teaches community members how to prepare and respond to disasters and take action until formal emergency response arrives. When Larysa began meeting with her local CERT, she found a place where her disaster-prepping tendencies could be put to good use. She learned how to assess the danger of a situation, put out small fires, do basic search and rescue, triage injured victims, and provide first aid, and she also learned more ways to prepare her home for hazards. Her isolated, late-night doomscrolling and internet-fueled apocalyptic panics subsided once she began to channel her anxious energy into hands-on activities with her neighbors.

BRINGING HEALTHY EMOTIONAL ENERGY
TO DISASTER RESPONSE

The Finns have a word, *sisu*, for the determination to see a task through. It is kind of like the quality of "grit," which positive psychology defines as a passion and perseverance in pursuing long-term goals and which is associated with success at things like sports, academic work, building projects, starting businesses, and creative pursuits.[13] The emotional regulation skills we have been learning can help us nurture and channel grit, or *sisu*, as part of healthy and sustainable disaster preparation.

Look back at the feelings list we used in Chapter 2 and imagine you are preparing for a disaster that might affect your home and family. First ask yourself: What am I feeling about all of this? And then: What are the sensations in my body? What actions are these feelings calling for? Then, ask yourself: What do I want to be feeling? What emotional energy do I want to bring to my disaster response? If I could feel anything I wanted, what would it be? Keep in mind that some of these may be *stretch feelings* and you will have to start with a middle-ground feeling that is on the way toward your desired emotional state.

Finally, ask yourself: What is the right thing for me to be feeling right now? What does the situation call for? Or in other words, what would be an adaptive level of eco-anxiety? Regulating your emotions will help keep you present and focused, allowing you to use your thinking and stress management skills to work effectively with others on disaster preparedness and response.

SHRINKING THE DISASTER

An essential "how-to-think" skill for disaster coping is being able to shift your attention from long-term, global-scale climate issues to more immediate local ones. When barreling toward an iceberg in a *Titanic* moment, you don't want to be focusing on the whole ocean.

Let's call this process *shrinking the disaster*. When I say shrinking, I don't mean minimizing its scope or danger, but rather putting it into focus and letting other, less urgent concerns shift to the side. Shrinking the disaster can give you a sense of control and help you determine priorities for problem-solving. It is a helpful antidote to the *catastrophizing* that often happens when we are faced with an unclear threat: when we exacerbate our distress by dwelling on worst-case scenarios and allowing the negative possibilities and what-ifs to grow in magnitude in our minds. Ironically, in a disaster caused by global climate breakdown, your world becomes hyperlocal. And, as Larysa reminded the newcomers to her CERT group, there can be good moments and good days even during a perilous time, like the sun shining through a break in the clouds.

DISASTER MENTAL HEALTH SKILLS

There is a photo I often show when teaching therapists how to deliver mental health services to people who have been impacted by climate disasters. It was taken in the aftermath of the Camp Fire near Chico, California, in 2018 and shows a man and woman with their backs to the camera, looking up at a large cluttered whiteboard at a FEMA disaster response center set up in a parking lot. The board contains neatly written lists of emergency services, information about shelters at local churches and schools, and tips on how to get help with relocation. There is a poignant "Looking For" section with handmade notices from people trying to locate friends and relatives. In the background are rows of tents, racks of clothing, and boxes of donated goods. We don't know much about the people in the photograph. We can't see their faces. The man has his arm over the woman's shoulder, suggesting they are a couple. The implication is that they have just lost everything except for each other.

Can you imagine what is going through their minds? How might they be feeling, physically and emotionally? How shocked, exhausted, wired, hungry, dirty, tired, sleepless, fearful, and concerned for loved ones must they be? If this were you, what would you need?

"Disaster mental health" is a specific framework therapists use to support people in the immediate aftermath of disasters and emergencies. It is short-term and practical, a kind of psychological first aid, or a distilled version of the process of stress management and "reclaiming your nervous system" we explored in Chapter 3. It involves providing those essential things that you might imagine the couple in the photo would need: to feel physically safe and emotionally calm, to be connected with loved ones and social supports, and to experience a sense of self-efficacy, control, and hope.[14]

Don't Create Another Victim

An old lesson from my wilderness first aid training has always stuck with me: "Stop and survey the scene; don't create another victim." In other words, don't inadvertently put yourself in harm's way or create more work for other rescuers. This adage is appropriate in many situations, including the aftermath of climate disasters, where we may feel compelled to help under dangerous or ambiguous conditions.

Be Patient

After the catastrophic Lahaina, Hawaii, wildfire of 2023, Victoria Liou-Johnson and other psychologists and professionals who were involved in the mental health response described how, in the immediate aftermath of the fires, it was important to become a temporary *disaster maven*: a point of contact for survivors, victims, and the many local and international helpers.[15] In a matter of days, the group organized a practitioner *hui* (group), facilitated a Zoom room for twenty-four-hour on-demand support and counseling, and trained and vetted the international cadre of mental health professionals who volunteered their services.

One of the major lessons they stressed for outside helpers was the value of patience. We've become so habituated to the instantaneous nature of news and information, ease of travel, and ability to communicate around the world in real time that we often forget what it looks like when these things break down. Paradoxically, at the disaster epicenter, often there is no communication. Systems we take for granted have gone down: roads and bridges destroyed, phone lines down, cell service and internet unavailable. In disasters, time slows down to survival in the present: walking or running to safety, digging out, locating survivors, finding shelter. It isn't safe to dispatch volunteers into action immediately. Donations have no place to go. By the time a system is created to coordinate help, attention is already shifting. But remember that once a disaster leaves the front page, the suffering continues. In fact, if you have the resources, one of the best times to help a community or region that has suffered a disaster is long after the relief efforts are over.

Training for Disaster Scenarios

While catastrophizing is an unhealthy and unproductive method of coping with disasters, confronting our fears and anxieties can, in moderation, help diffuse them. In Chapter 10 we learned how Larysa practiced imagining disaster anxiety triggers while simultaneously calling forth a relaxation response from her body. One reason that athletes, emergency responders, pilots, and soldiers react calmly and thoughtfully in high-stress situations is that they have visualized and anticipated these stressful scenarios many times before. That's why I want to get you to the place of "I already know" about climate threats.

Larysa knew she had made great strides in her mental preparedness when she read writer Kathryn Schultz's *New Yorker* story "What a Major Solar Storm Could Do to Our Planet," which described in terrifying detail how a major solar flare could be expected to paralyze satellites, GPS signals, and radio communications, devastating our global infrastructure and communication in unpredictable ways.[16] Nearly a decade earlier, Schultz's 2015 Pulitzer Prize–winning story "The Really Big One"—about the havoc a large Cascadia Subduction Zone earthquake would wreak on the Pacific Northwest—had been fuel for Larysa's eco-anxiety and sleepless nights.[17] This time however, instead of spending her nights worrying and ruminating, Larysa discussed the article with her local disaster preparedness group, and they incorporated this issue into their disaster response planning. She was no less aware of the seriousness of the issue than she would have been in her doomscrolling days. But her response was different. She could genuinely say "I already know," not about this rare solar flare issue, but about disaster response in general, and intuit ways she and her community would be able to cope.

———

The ability to mentally prepare for and adapt to disasters is a core attribute of environmental awareness in the modern world. If a climate disaster is your entry point to uncovering your environmental identity, then the crisis has given you an unexpected opportunity. But you don't have to wait for extreme situations to reassess your connections to Earth; you can get ahead of the storm surges to come. Becoming more knowledgeable and confident about disaster response is an important step toward developing the mindset of a *climate survivor*.

EPILOGUE: MASTERING TWO WORLDS

So the wise soul governing people would empty their minds,
fill their core, temper their ambition, strengthen their resolve
—Tao Te Ching[1]

THE FINAL HOMECOMING OF THE HERO'S JOURNEY TENDS to be overlooked in comparison to the drama of the quest and the trials encountered along the way.[2] This is the "master of two worlds" stage: when the adventurer has overcome obstacles in both the outer, physical world and achieved personal growth in their inner world. Having acquired the wisdom and skills to bridge the gap between the two, they return to their community with newfound gifts to share.[3]

If you think back to the exercise I shared at the very start of the book, with people expressing their feelings about the blue globe of Earth and the flaming globe, you'll recall that to "master" in this case means to learn how to *hold* the two worlds, like the globe props in the exercise. It means having the capacity to feel all the feelings, to be able to hold both the hope and the pain of living in this era of global climate breakdown. This is what it means to be a *climate survivor*, someone who has overcome obstacles, has acquired new wisdom and skills, and is ready to share those gifts with others.

Your gift to share with your people and place is a resilience that allows you to mourn your losses, adapt to your present reality, and trust in your hopeful vision for the future of Earth and human society that extends beyond your lifetime. In this book, I have witnessed and shared in your journey. Now is your time to share in and witness the journey for others.

Another overlooked aspect of the hero's journey is that while we may seek these stories as entertainment, they are not pleasant for those who will live them. The quester is forced to endure trials, often against their will, and often in the face of great danger. To get out in one piece, one can neither become hostage to despair nor become enchanted by utopian visions.

Inevitably, coming home leads to another beginning. The journey to becoming a climate survivor is not a straight path, but a cycle. Though you have reached the end of our time together, your journey is not over. Rather, it is "to be continued." Your sense of doom will come and go. You will face more stressors and more trials. You'll realize you need to check in with yourself and ask: "How am I thinking, feeling, managing stress, dealing with direct and indirect climate impacts? What are my values? How can I deepen my relationships and interbeing with nature, and know and protect my standing place? How can I congregate with my community and my eco-friends? How can I be generative and help others? And when is it time to rest and recharge?" Remember, despair is often fatigue in disguise.

With each cycle, you will acquire new tools, new knowledge, and new ways of flourishing. The same is true of the characters you have met throughout this book. Larysa went from late-night doomscroller to community emergency response leader. Sally learned how to better communicate and compromise with his partner, while still standing up for his values. The outdoor adventure Christina thought impossible slowly became possible when she committed to her pilgrimage. Cancer reminded Liberty of her mortality and made her eco-values even more clear. Braden found a way to tell a sustainability story of timber-framed skyscrapers and small-town survival. Marcus learned that if you go back far enough in your ancestry you will find someone who knew how to live on the land, and it's never too late to recover those traditions, no matter where you are from or the color of your skin. James learned to trust that he could take a sad song and make it better.

As for me, what I've discovered through the process of writing this book is that what we look at defines us. When we dwell on climate catastrophe, it becomes all we see. At the same time, when we stare long enough into possibilities for flourishing, we find those possibilities staring back at us. It's not that we're blinded to all of the problems. It's just that the problems aren't dominating our field of vision. Problems are defined more clearly when surrounded by good things. We can see their borders and edges. They don't seem so insoluble.

I don't know what your future will look like. But I do know that there will be good days. Days when your horizon of hope will expand beyond what you can see and become faith. Other times it will contract so painfully you'll have to remind yourself to breathe. And then you'll get back to the basics: "What am I feeling, what do I want to feel, what do I need to feel?" It is my hope that you'll then draw on all that you have learned in these pages, and have faith that even when the fate of the world feels the most dire, you have the tools to cope, heal, and even thrive.

Meanwhile, I'll be here in my place, tending the fire. And you'll be in my thoughts and heart. For now, safe home, blessings to you, and may you walk in beauty. Fare well.

ACKNOWLEDGMENTS

To the human and the more-than-human world, and our shared future. Specifically, I would like to thank my clients, teachers, colleagues, friends, and family. This manuscript would not exist, or be shared so widely, without the hard work and dedication of my collaborative editor, Lilly Golden; my agent, Laura Nolan; and Talia Krohn, Jules Horbachevsky, Jessica Chun, and the team at Little Brown, Spark. Tracey Behar provided crucial early support. Again, to all, warm thanks!

APPENDIX: HOW TO TALK TO YOUR THERAPIST ABOUT CLIMATE CHANGE

Therapists and counselors with a specialty in eco and climate concerns are still relatively rare. However, one of the best predictors of successful outcomes in counseling is the fit between the client and healer, regardless of specialty. There are so many kinds of therapy, and if you're new it can be bewildering. I'll give you some tips.

Your engagement with counseling or therapy depends on the depth and ambition of your goals: to work through injuries and losses, putting them into perspective; to learn and practice new life skills; or to change your emotional expression and communication patterns. Even if you can't change an issue or reality in your life, you can change how you think and feel about it, and how you approach or adapt to it. You don't need to have major mental health issues before you seek therapy. Some folks like to use counseling or therapy as an ongoing practice, a basic part of their health care.

LEVELS OF ECO AND CLIMATE THERAPY

The many levels of therapy that you might seek can include:
1. Basic coping using the provider's therapy framework and tools
2. Support for identity development (including environmental identity) and for personal growth
3. Help with care and healing of severe symptoms, concerns, diagnoses

4. Coaching or consultation on performance, wellness, thriving
5. Help with taking action on personal health or sustainability-related behavior changes
6. Comprehensive approach to the issues in this book

SCRIPTS

Here are a few things you can say when you reach out to a mental health practitioner about your eco and climate concerns.

As a New Client

Hello, I am seeking a provider to help me apply mental health, counseling, therapy, or coaching to address my concerns about climate and environmental issues. I'm seeking a provider who has some familiarity or training in this issue, or who is willing to use their existing expertise to help me with some of the symptoms and discomfort I am feeling. I don't require that the provider be an expert on climate or environmental issues, but simply that they recognize these are real and pressing issues in themselves, with demonstrable health impacts, and are willing to help me work on those.

To Use with Current Therapist, Counselor, or Coach

I would like to expand the focus of what we are working on to include my thoughts, feelings, concerns, and/or impacts that I am experiencing about environmental issues and climate change. These could be specific events either local or worldwide, or concerns I have about the future or making life decisions about relationships, having children, career, where to live, and so on. I see these environmental concerns as being on par with some of the other things that we have worked on, and I would like to apply or adapt some of the skills and benefits I've already received in this direction.

SPECIFIC NEEDS AND LIFE CONTEXTS

People who might need to address climate concerns in therapy:

- Those having an environmental consciousness-raising ("waking-up syndrome")
- Disaster victims and trauma survivors
- People with preexisting mental health issues or diagnoses made worse by eco and climate problems
- People facing major life decisions or transitions (job, career, family, where to live)
- "Climate workers" such as public health and safety professionals; elected officials and government workers; scientists, researchers, and journalists; social and environmental activists

In all stages of life, therapy may also address concerns about climate:

- *Children,* who may have age-appropriate anxieties, have temperament issues, be precocious learners who confront troubling information, or have experienced negative effects from disasters or migration
- *Adolescents,* including students, high-functioning teens taking on leadership and advocacy, youth having identity development challenges or life stressors
- Young *adults,* particularly those immersed in activism, forming new identities, or creating a new life outside of their families
- *New or prospective parents or grandparents*—those making childbearing choices, having concerns for their children, or experiencing fatigue and stress from parenting or supporting earlier generations
- *Eco or climate professionals,* including those working in the sciences, conservation, and public health, and including issues with work-life balance, maintaining healthy relationships, thinking style and emotional intelligence and expression, or severe fatigue and burnout
- *Elders,* especially those facing health and mortality concerns, concerns about younger generations, dealing with impaired family relations, or experiencing despair or feelings of failure

See the *Personal Sustainability Assessment* below for an additional way to think about the many dimensions of your well-being.

HEALTH INSURANCE

Health insurance generally does not cover any therapy unless it is in relation to an identified medical health need. That is, insurance usually will not provide coverage if the reason for the therapy is personal growth, couples relationships, or eco-therapy. To utilize these benefits it is best to focus on the specific impairment or symptoms (e.g., feelings of anxiety, depression, adjusting to stressors, dealing with disaster traumas).

SPECIALTIES

Some specialties to ask about include outdoor or walking therapy, disaster recovery, specific approaches to eco or climate therapy, couples and families, children and adolescents, trauma, racial and cultural competency, familiarity with LGBTQ+ life, openness to recognizing social justice issues, and the ability to address spiritual concerns or align with your faith tradition. Further specialties or needs could be chemical sensitivities, science and environmental professionals, military, veterans, and first responders.

LOCATING A PROVIDER

As far as resources for locating a provider, you can consult some specialized directories sponsored by the Climate Psychology Alliance North America or the Climate Psychiatry Alliance. You can also search for providers who describe themselves as "climate conscious" or who include environmental issues or climate change in the list of issues they address. Otherwise, you can find providers who are prepared to address your chief complaint (again, the impairments or impacts, such as mood or sleep issues, anxiety and panic, severe grief and despair) and seek help with those.

YOUR PERSONAL
SUSTAINABILITY ASSESSMENT

WHEN I WAS GETTING STARTED AS A PSYCHOLOGIST, I looked for ways to help clients and therapists seeking a way to add an environmental focus to their personal and social well-being. This led to the idea of *personal sustainability*—applying environmental sustainability principles to your own health and well-being, such as with rest, diet, and social support. Then I created self-assessments for people to determine the priorities of a healthy life from this perspective. Here's a simple version.

All the areas below are potentially important to your well-being. As you consider them, notice which might be missing or lacking attention. Perhaps you didn't realize you could make this an area of well-being. Also, and most importantly, notice where there are discrepancies between how much you value or prioritize one of the dimensions and how satisfied you are with your functioning or progress in that domain. At the end of the assessment, you can use your findings and the questions that arise to guide your own personal sustainability development, or to seek help and support.

Your Personal Health
- *Thoughts.* Focused and effective in your thinking, clarity about your beliefs and values.
- *Emotions.* Experiencing and expressing a range of feelings, and not feeling blocked.
- *Body.* Physical functioning, fitness, feeling of energy and wellness.

- *Diet.* Satisfaction with food choices, eating habits, cooking, and related health and ethics, and/or being involved in food systems or production.
- *Happiness with life.* Enjoyment and having fun.
- *Creativity and play.* Making or creating, arts or crafts, having a hobby.

Relationships

- *Social life.* Being accepted by a group(s), being assertive, having your needs met.
- *Friendships.* Finding and maintaining long-term close and friendly relationships.
- *Romance / significant other.* Having a safe, trusting, and satisfying intimate relationship.
- *Sexuality.* Having a safe and enjoyable sexual life and identity.
- *Parenting / mentoring.* Caring for and supporting the needs of children and younger generations, and/or being a parent (biological or nonbiological), mentor, or caregiver.

In the Community

- *Livelihood.* Having meaningful work, providing for needs, making ends meet.
- *Career, business, or vocation.* Working to potential, fulfilling tasks, performance, and success.
- *Civic life.* Engaged and empowered as a citizen, being active in politics, community service, or social and environmental justice (local or global).

Environment and Sustainability

- *Connection with nature.* Knowledge of and comfort with the outdoors and other species; time spent in nature for work, health, or recreation.
- *Sustainability.* Comfort with lifestyle, level of consumption, and ecological or carbon footprint.

- *Adaptation to climate change.* Being prepared for weather changes and disaster threats.

Wisdom and Transcendence

- *Maturity.* Living appropriately for your stage of life, a sense of being on the right track.
- *Service.* Helping others, making a positive difference in the world.
- *Religion / spirituality / soul.* Finding meaning, feeling connected to something larger than yourself; being part of a faith tradition; being at peace.

Now it's time to review. What were the most important areas for you? What areas emerged as having the most—or the least—satisfaction? Do you see any potential next steps? What aspects of a healthy life would you add? How might you prioritize areas you want to work on or improve? What information or support do you need to move forward? (See "How to Talk to Your Therapist About Climate Change" above.) And finally, how does it feel to look at your life from the perspective of personal sustainability?

GLOSSARY

Adaptation: In terms of climate breakdown, taking action to safeguard oneself and community to create a good life in the face of changing weathers and disaster threats; making choices to live in concert with your values; helping other people and beings; and finding positive opportunities within the crisis.

Affordance: How your perceptions and capabilities fit with the possibilities the physical environment offers; for example, when climbing a tree, a branch feels right for your hand because your hand evolved to grasp the branch.

Alexithymia: Lack of language or vocabulary for discerning and expressing feelings, or for recognizing the feelings of others; can be a barrier to coping with eco and climate distress.

Anthropocene: A time of rapid and at times unpredictable geologic events prompted by widespread human transformation of Earth's natural systems.

Assimilation / Accommodation: Two ways of learning associated within growing environmental awareness. Assimilation involves fitting new information into an existing mental model, and accommodation involves creating new categories and meanings to make sense of novel information.

Black to the Land: A program and movement to restore nature connections, outdoor recreation, farming, hunting, and gathering among African Americans.

Broaden and build: A positive psychology approach to emotions and well-being based on cultivating uplifting feelings to foster coping, creativity, and growth.

Capacity: Ability to mindfully hold awareness of issues, your emotions, and impulses, and to operate in relation to your values and goals. See also *psychological flexibility.*

Capital-I issues vs. little-i issues: Capital-*I* issues are the important and large-scale social, political, cultural, and environmental matters that motivate you. Little-*i* issues are your individual attributes, gifts, and challenges, including personality, style, strengths, weaknesses, wounds and vulnerabilities, and relationship dynamics.

Carbon fee and dividend system: Also called climate income, a system to reduce greenhouse gas emissions that taxes the sale of fossil fuels and distributes the revenue of this to the entire population (equally, on a per-person basis) as a monthly income or regular payment; currently used in Canada, Switzerland, and Austria.

Carbon market: A system through which countries, organizations, and people can buy and sell permits to either produce or remove carbon dioxide emissions and other pollutants from the atmosphere.

Carbon offset: A climate-friendly activity done in one place in the world (like planting trees or employing carbon capture technologies) that will remove an equivalent amount of carbon emissions created somewhere else.

Celsius vs. Fahrenheit: Average global heating is measured in degrees Celsius above preindustrial levels. Readers in the United States need to keep in mind that each degree Celsius is nearly double (more specifically, 1.8 times larger than) each degree Fahrenheit, so a 2°C increase would be equal to 3.6°F of heating, a 3°C increase would be 5.4°F, etc.

Citizens' Climate Lobby: A grassroots group that helps volunteers build relationships with their elected representatives to create political support across party lines for putting a price on carbon, like by using a carbon fee and dividend system at the national level.

Citizen solar movement: Grassroots groups that advocate for solar power options at the local and state level—for example, the American Solar Energy Society, Save California Solar, and Vote Solar.

Climate breakdown: The collective effects of harmful, extreme shifts in global temperature, weather, and other environmental changes caused by the accelerated burning of fossil fuels and other human activities; especially urgent in regard to public health and societal collapse.

Climate elephant / blind people and the elephant problem: In relation to climate breakdown: when we reduce complex, systemic situations and problems to a single manifestation or solution that makes the most sense to us, and are blind to other ways people see and seek to solve the problem.

Climate hostage: An individual experiencing a sense of entrapment and peril given the scope of eco and climate threats, the lack of effective response by leaders or economic systems, and the limits of their individual efforts.

Climate migrants / refugees: People compelled to take leave of or flee their homes or traditional lands due to environmental disasters or gradual degradation of their ecosystem.

Climate Psychology Alliance: Network of mental health providers (UK, US, and international) committed to addressing the impacts and causes of global climate breakdown and supporting the public through effective therapeutic response and advocacy; a resource for locating a climate aware counselor or therapist.

Climate survivor: A self-chosen term for someone who actively takes on healthy living in the context of climate breakdown, as well as the tasks of well-being, self-protection, and finding meaning. A mindset that goes beyond seeing climate change as a catastrophe and tragedy, including possibilities for coping, opportunities for growth, positive societal changes, and thriving in a new world. See also *Adaptation, Post-doom*.

Climate vs. weather: Climate is an idea, a scientific and cultural concept, the long-standing averages in temperatures, precipitation, and vegetation in a place or region. Weather is a physical reality you can touch and feel, the short-term atmospheric conditions in a specific place. Rapid breakdown of the familiar climate conditions of the last 10,000 years caused by global heating is creating many different and unpredictable local weathers. See also *Anthropocene*.

Community Emergency Response Team (CERT) program: A program that trains and supports local volunteer responders to assist in fire safety, basic search and rescue, and medical and psychological first aid during disasters. Also includes Neighborhood Emergency Teams (NET) and other regional variations.

Dark ecology/ bright ecology: Large systems exhibit a foreboding dark ecology when you perceive many interconnected health threats. Simultaneously, you can encounter an inspiring bright ecology when you discover solutions, innovations, supportive allies, and new ways to understand yourself. See also ***hyperobjects.***

Deep ecology: A holistic philosophy about environmental science and advocacy that recognizes the inherent worth of all beings and natural systems beyond the benefits to humanity and human uses. Contrasts with so-called shallow or reductionist (human-centered) views of ecology.

Doomscrolling: Compulsive online searching regarding threats, disasters, and other distressing content; can be motivated by curiosity or a duty to be informed or to bear witness, as well as anxiety, fascination, and escapism; often amplified by attention-driven algorithms favoring content that increases perceptions of threat and distress.

Eco-distress / eco-anxiety: Troubled feelings of anxiety, sorrow, or pain related to problems or threats in our human relationship to nature. Often experienced in the form of fearful or debilitating feelings of threat, such as from toxins in the environment or impending disasters; at other times, manifesting as loss, grief, and despair.

Ecological: Pertaining to interrelationships between and among all life-forms (including humans) and natural systems, at levels of scale from the microscopic to the ecosystem, planet, and beyond.

Eco-timeline: An exercise to trace your experiences of nature and the development of environmental beliefs, values, and identity through your life; prepares you to live your identity and values more consciously in the future.

Empathic distress: Being caught up in and overwhelmed by the distress of other beings, amplified by the global nature of eco and climate problems, and over time leading to emotional depletion and exhaustion.

Environmental identity: Your self-concept, beliefs, and values about your being and relationship to the natural world and other species; intersects with other forms of identity and diversity such as race, class, culture, ethnicity, gender, sexual orientation, and ability.

Food sovereignty: Control, self-determination, and self-sufficiency regarding one's food supply.

Forest bathing: In Japanese, *shinrin-yoku*, a therapeutic practice of immersing oneself in the sights, sounds, and scents occurring in forested natural settings; used to promote relaxation and immune response; often fostered by protocols of rest and sensory experiences, and by using dedicated parks and walking paths.

Galveston storm of 1900: An unexpected hurricane and storm surge that struck the low-lying city of Galveston, built on a sandy barrier island, on September 8, 1900, resulting in the deaths of 8,000 or more people in under twenty-four hours, with thousands more injured or made homeless; the most deadly natural disaster in US history, and a benchmark for the severity of tropical storms.

Gates rule / Moore's law corollary: The observation that technology (automation, industrialization, digitization) applied to an efficient process will tend to magnify the efficiency, while technology applied to an inefficient process will magnify the inefficiency; further, the pace of either positive or negative effects will tend to increase exponentially given increasing computing power and other factors; both observations are applicable to social and technological forces associated with out-of-control climate breakdown.

Grief map: A way of illustrating environmental losses, such as extinctions and damaged life ways, experienced in the past or the present or anticipated for the future; a tool for expression, meaning-making, and identifying opportunities for proactive efforts to prevent future losses.

Growth mindset: A belief that your talents and intelligence can be developed through effort and practice, versus being fixed and innate; associated with persistence, growth, and improved performance over time; applicable to your beliefs about finding solutions to eco and climate problems.

Harm reduction: An approach to health care that seeks to limit the negative effects of issues and behaviors even if the issues or behaviors themselves cannot be eliminated; often based on public education and empowerment; can be applied in the context of coping with chronic eco and climate problems.

Homo aequilībrium: A perspective on humanity not based on the use of tools or knowledge; seeking balance in terms of mind and human and natural systems; a response to the challenges of polycrises; can begin with personal sustainability.

Honest broker: An approach to science policy advising that focuses on highlighting the outcomes of different alternatives and approaches and letting the policymaker decide their actions; contrast with value-free basic science or being an issue advocate; can be applied in an ecotherapy context.

Hyperobject: An event or phenomenon that is massive in space and time relative to human experience and understanding, often giving us a sense of it being simultaneously *beyond* us and *inside* us, such that it requires new ways of perceiving to understand or interact with.

Imposter syndrome: A belief you are inadequate to a task, or that your abilities are misjudged or exaggerated by others; often privately held and leading to feelings of shame; often operates when people take on complex eco and climate issues.

Land acknowledgment: A formal statement recognizing the sovereignty and rights of Indigenous or First Nations peoples of a place or territory; as part of environmental identity development, a process of coming into conscious relationship with and stewardship of a local place and ecosystem, and all its beings and peoples.

Little-i issues: Your individual attributes, gifts, and challenges, including personality, style, strengths, weaknesses, wounds and vulnerabilities, and relationship dynamics

Mass timber: An engineered wood product that is strong enough to build large structures like high-rises and airports; valued for its lower carbon footprint, versatility, and beauty.

Master recycler program / Master gardener program: Community volunteer corps supported by states and universities focused on sustainable consumption, repair, reuse, composting, and recycling, and on cultivating healthy communities through sustainable horticulture education and gardening projects.

Minari: Korean wild green cultivated for food, and transplantable to new lands; a 2020 film about a South Korean family living in the rural US South by Lee Isaac Chung; a metaphor for becoming rooted in a new place for Korean and other Asian people.

Mindfulness: The process of paying attention, on purpose, to the present moment while withholding judgment; associated with many aspects of personal health, stable mood, insight, and growth

More-than-human world: The physical world of places and species that exists before and outside of human affairs, with its own relevance and agency, and that we can enjoin, share kinship, or serve; in contrast with usages like "nonhuman" that imply nature is alien, possibly dangerous, lacks moral standing, or is an expendable resource. See also *deep ecology*.

Nearby nature: Green spaces and organic processes local to your home, neighborhood, or community.

News fast: Taking a break from exposure to journalism, commentary, and information gathering, particularly through electronic or online media, to lower stress, find mental and physical restoration, and regain perspective about the living world and relationships around you.

Outrage fatigue: State of exhaustion caused by chronic exposure to distressing news and information, particularly attention-grabbing, algorithmically driven content; also related to overreliance on a limited spectrum of emotions related to moral judgment and self-righteousness.

Overview effect: Sense of awe, shared humanity, kinship, and inspiration associated with viewing Earth from above, as in orbit or from space; perception of the planet as a discrete and self-contained system.

Personal sustainability: Applying environmental sustainability principles to your own health and well-being, such as rest, diet, and social support; having a sense of continuity and resiliency in one's life, relationships, and work.

PFAS: "forever chemicals," human-made per- and polyfluoroalkyl substances used in industry and consumer products that persist and accumulate in the environment, causing decreased fertility, birth defects, and cancers among humans and other species.

Post-doom: A stage of coping and adjustment regarding climate and environmental issues that follows consciousness-raising and disillusionment, characterized by experiments in new forms of being, feeling, and action.

Precarity: A chronic state of financial insecurity regarding meeting basic needs (shelter, food, and personal and family development); exacerbates the effects of environmental disasters; associated with government and economic policies.

Present-moment safety: The ability to recover a sense of assurance, security, or refuge in the here and now while holding the awareness of dangers and threats.

Psychological flexibility: Ability to cope with, accept, and adjust to difficult situations, and to act in line with your values and long-term goals; a quality important in an eco and climate context.

Resilience: A widely applied concept in culture, mental health care, and disaster response, referring to the ability to adapt well to difficult situations and bounce back from adversity; called "transformation resilience" when encompassing a new or improved state; also used to describe durability and resistance to change in human and natural systems.

Restorative-familiar / diminutive / deep flow: Qualities of restorative natural environments. *Restorative-familiar* refers to the pleasure of revisiting a place one knows well; *diminutive* refers to a sense of awe and smallness in the face of nature's scale and grandeur; *deep flow* refers to absorption and focus in the rhythms of an outdoor place or activity.

Rewilding: Philosophy of ecological restoration that encourages biodiversity and promotion of natural processes and species inherent to the place; a movement toward reverence for wild processes and places in human health and culture.

Scale vertigo: A sense of disorientation when confronting the size and extent of eco and climate issues.

Sense of place: Qualities of experience, familiarity, belonging, emotional connection, and security in relation to one's home landscape, community, or territory; may be a given or can be developed and deepened.

Shifting baselines: Differences and blind spots in how succeeding generations perceive what is normal for the integrity and biodiversity of the natural environment; lack of knowledge of the more pristine conditions of the past can increase tolerance for degraded environments and lower expectations for protections and restoration.

Solastalgia: Loss of comfort and ease as your local natural environment becomes damaged and transformed, such as through open pit mining, persistent drought, wildfire, or other climate and environmental changes.

Somewhere ideas / nowhere ideas: A *somewhere idea* is a program or action that is practically applicable in a real, local place and community; contrast with a *nowhere idea*, an abstract idea that is not situated in local context. See also **land acknowledgement**.

Standing place: Adapted from the Maori concept of *turangawaewae*, literally "a place to stand," an ancestral or adopted place where you situate yourself and identity, and that you join with others to revere and protect. See also **sense of place, environmental identity**.

Strategic underachievement: A technique to manage one's time and priorities by identifying areas of your life and work where you will expect good-enough but not excellent performance, so you can devote your finite resources to higher-priority goals; useful for cultivating personal sustainability and environmental action.

Sunrise Movement: A grassroots group that advocates political action to address climate breakdown such as policies to scale up the use of renewable energy, create livable wage jobs, and promote social justice; traditionally has a young, politically progressive membership.

Sustainable consumption: A philosophy of choosing, using, and disposing of products in a way that minimizes human and environmental impacts; attention to the life cycle of products, including sourcing, production, carbon emissions, workers' conditions, durability, potential for reuse / recycling, and safe disposal and degradability.

Thwaites Glacier: The so-called Doomsday Glacier, an immense West Antarctic glacier that has achieved global prominence given its precarious state due to global heating and its potential to spur catastrophic sea level rise in the event of its collapse.

Unnatural disasters: Unprecedented, catastrophic events such as storms or wildfires that are unusually large or occur in novel places in comparison with historical experience; calamities where human actions or infrastructure failures worsen the impact of naturally occurring hazards.

Un-solo: Variation on the theme of a solitary wilderness or outdoor camping retreat, with the implication that the person is not alone but is accompanied by the place, local species, weather, and season.

Upside-down pyramid: The perception of inadequate personal resources to cope with all the social and environmental stressors a person is experiencing.

Validate, elevate, create: A counseling and coaching technique that explicitly recognizes the appropriateness and importance of a person's concerns and encourages novel or inventive thinking regarding next steps or solutions; applicable to eco and climate concerns that are often dismissed or minimized.

Values-to-action pathway: A theoretical perspective on how your environmental values interact with your knowledge, beliefs, and level of self-efficacy in fostering a motivation toward environmental action.

Waking-up syndrome: An unsettling or abrupt conscious raising about the seriousness of climate problems, often linked to a discrete life experience or epiphany.

Wicked problems / wicked solutions: A *wicked problem* is a complex, multi-issue, high-stakes problem that defies simple definitions or linear solutions and has a high risk of producing damaging outcomes. A *wicked solution* is a commensurate response that includes patient engagement, understanding, and novel, non-obvious solutions designed to minimize unintended harm.

Wilderness: A landscape that has not been modified by human activity and retains its self-organizing ecosystems and character, or is managed to be so through a legal designation as a wilderness or scenic area; an area perceived to be remote, rugged, or inhospitable that presents opportunities for escape, solitude, and challenge.

Wildland-urban interface: Zones where homes, towns, and metropolitan areas meet and intermingle with forests, mountain regions, grasslands, and other natural vegetation; areas at high risk for devastation from wildfires and other calamities exacerbated by climate breakdown.

SURVIVING CLIMATE ANXIETY: THE OFFICIAL PLAYLIST

Scan the QR Code for a playlist designed to go along with the book. The playlist captures the essence of *Surviving Climate Anxiety* through a blend of classic and contemporary songs and spoken word, each selected to reflect different facets of environmental identity and the emotions evoked by the climate and ecological issues we are facing. From poignant to playful to impassioned, the songs weave together a tapestry of sounds that can inspire you on your own journey.

The wide-ranging approach to the playlist is inspired by "climate playlists" shared by Panu Pihkala and me over our years of collaborating on the *Climate Change and Happiness* podcast (see Chapter 15 for details).

RESOURCES AND RECOMMENDED READING

Anyone who has a book collection and a garden wants for nothing.

—CICERO

Books, resources, groups, and initiatives all occupy parts of the climate elephant. Many have competing visions, especially about how best to employ technology, organize the economy, and address historical wrongs. As a climate cosmopolitan, it's good for you to tour them and see for yourself what is being thought of and done. This will expand your worldview and your environmental identity. You might be surprised by all the good-faith efforts on the part of your fellow humans. So don't just learn, celebrate. Use the lists below to get started.

Resources

I recommend finding a group of resources and causes local to you that celebrate and protect all life. My own recipe includes music, family, places, and species local to where I live in the Pacific Northwest of the United States:

- KEXP (a local and international source of new and classic independent music): https://www.kexp.org
- Families for Climate: https://www.familiesforclimate.org
- Human Access Project (ensuring safe access to the Willamette River): https://humanaccessproject.com/
- Elakha Alliance (restoring the Pacific sea otter ecosystem): https://www.elakhaalliance.org

Get Involved

- Families for Climate: https://www.familiesforclimate.org
- Citizens' Climate Lobby: https://citizensclimatelobby.org
- Braver Angels: https://braverangels.org
- Black to the Land: https://www.blacktothelandcoalition.com
- Anthropocene Alliance: https://anthropocenealliance.org
- Outdoor Afro: https://outdoorafro.org
- Hispanic Access Foundation: https://hispanicaccess.org
- Venture Out Project (queer and transgender community): https://www.ventureoutproject.com
- Adaptive Adventures (bringing progressive outdoor adventures to individuals with physical disabilities and their families): https://adaptiveadventures.org

Create and Celebrate

- In a Landscape: Classical Music in the Wild: https://inaland scape.org
- Music Declares Emergency: https://www.musicdeclares.net
- Invisible Dust (blending art and science): https://invisibledust.com
- Gallery Climate Coalition: https://galleryclimatecoalition.org
- The International League of Conservation Photographers: https://www.conservationphotographers.org
- The Climate Museum: https://www.climatemuseum.org/

Learn Different Ways to Make Change

- 80,000 Hours (careers that solve pressing problems): https://80000hours.org
- Tools of Change (for designing behavior change programs): https://www.toolsofchange.com/en/tools-of-change
- Rocky Mountain Institute (resources for the clean energy transition): https://rmi.org
- Breakthrough Institute (harnessing technology and human ingenuity for good): https://thebreakthrough.org
- Network for Engineering with Nature: https://n-ewn.org
- Alliance for Just Deliberation on Solar Geoengineering: https://sgdeliberation.org/about/mission-and-principles

Health and Thriving

- Climate and Mental Health Network: https://www.climatemental health.net
- Child and Nature Network: https://www.childrenandnature.org
- The Good Grief Network: https://www.goodgriefnetwork.org
- Climate Psychology Alliance (for mental health professionals): https://www.climatepsychology.us (North America and UK), https://www.climatepsychologyalliance.org
- Health Care Climate Council (for hospitals and medical professionals): https://climatecouncil.noharm.org
- The Yale Forum on Religion and Ecology: https://fore.yale.edu
- Susty Vibes (supporting young people in Africa): https://susty vibes.org

Climate and Environmental Science

- The Yale and George Mason Centers for Climate Communication: https://climatecommunication.yale.edu; https://www.climatechangecommunication.org
- World Weather Attribution Initiative: https://www.worldweatherattribution.org
- International Panel on the Information Environment: https://www.ipie.info
- Stockholm Resilience Centre: https://www.stockholmresilience.org/about-us/our-story.html

Adaptation and Disaster Response: United States*

- The US Federal Disaster Distress Helpline can be reached by calling 1-800-985-5990 and is reachable 24 hours a day, staffed by multilingual counselors at call centers across the country.
- The National Climate Assessment: https://nca2023.globalchange.gov
- National Disaster Recovery Framework: https://www.fema.gov/emergency-managers/national-preparedness/frameworks/recovery

---

* At the time of writing, some US government websites may be unreliable due to censorship and defunding.

- US Critical Infrastructure Sectors: https://www.cisa.gov/topics /critical-infrastructure-security-and-resilience/critical-infra structure-sectors

Your Reading List

To inspire you to do your own deeper dive, I'll recommend some of my personal library of readings that are meaningful. Key books, alongside music, paintings, sculptures, or poems, are building blocks of your environmental identity and landmarks on your eco-timeline. They help you create a room of one's own—or liberate an open space—for your sensibility. Some readings are evergreen, classics we all can return to again and again for new facets of their wisdom. Others reveal worlds being born or yet to come. In good times, they are treasures you can hoard like a mythical dragon or share like a philanthropist. In hard times, you can turn to them for support and to remind you of your values and meaning. To echo poet T. S. Eliot, these are fragments shored against our ruin.

A reading list is not just for filling your head with facts and information, and certainly not a contest to keep up on all the news and every publication. Strike a balance as to breadth. Yes, it's okay to specialize in one part of the climate elephant (one leg, such as scaling up clean energy use or adaptation to severe new weathers). But beware of silos, and don't read so narrowly that you lose a sense of the whole or of areas that you may not see yet.

How I Learned to Understand the World

There are some texts that are foundational to my understanding:

- Shierry Weber Nicholson's *The Love of Nature at the End of the World.*
- The Sierra Club anthology *Ecopsychology.*
- Mitch Thomashow's *Ecological Identity.*
- Neil Evernden's *The Natural Alien.*
- David Abram's *The Spell of the Sensuous.*
- Bill Plotkin's *Nature and the Human Soul.*
- The conservation psychology research and writings of Susan Clayton and Carol Saunders provide an essential empirical basis for environmental identity and for ecotherapy. See *The Oxford Handbook of Environmental and Conservation Psychology.*

- *The Transition Handbook* and the work of Rob Hopkins and others in the Transition Town Movement.
- Mike Hulme's *Why We Disagree About Climate Change* helped me to understand the meanings of climate and the possibility of a psychology of climate change.
- For anyone concerned about the oceans, or living along the North Atlantic coast, W. Jeffrey Bolster's *The Mortal Sea* is required reading.
- To gain a perspective of the Global South, see Amitav Ghosh, *The Great Derangement*.
- Paul Slovic helped me to understand why people become numb to large numbers in relation to disasters and causalities, see *Numbers and Nerves*.
- As someone of western European heritage, I found Simon Schama's *Landscape and Memory* a key resource for understanding the Indigenous identity underlying the landscape of Europe, as was Ralph Metzner's *The Well of Remembrance* on northern European earth myths.
- As an inhabitant of the American West, I found works like those of philosopher Viola Cordova essential in gaining a sense of the Native American worldview (see Cordova's collection *How It Is*).
- I owe William Loren Katz's *The Black West* for my appreciation of the presence of African Americans on the frontier, and William Kittredge's *Owning It All* for understanding the ranching heritage in the high desert of eastern Oregon.
- Terry Tempest Williams's *Refuge* taught me about being a downwinder, and Marc Reisner's *Cadillac Desert* taught me about rivers, water, and the landscape.
- I learned early on that what we call environmental thought and sensibility is not new; see William Wordsworth's 1810 *A Guide Through the District of the Lakes*, George Perkins Marsh's 1850 *Man and Nature*, and the wide-ranging works of Gene Stratton-Porter (begin with her 1924 *Good Housekeeping* story "The Last Passenger Pigeon").
- For understanding the more-than-human Earth and other

species, see Ed Yong's *An Immense World*, Zoë Schlanger's *The Light Eaters*, and Ferris Jabr's *Becoming Earth*.

Self-Help
While developing our environmental awareness and identity, we all can benefit from some solid self-help references. The well-worn and much-shared copies on my shelf include:

- Carol Dweck's *Mindset*.
- Alex Korb's *The Upward Spiral* (on depression and anxiety).
- Bessel Van der Kolk's *The Body Keeps the Score*.
- Thelma Bryant's *Homecoming*.
- Rick Hanson's *Hardwiring Happiness*.
- Chris Johnstone's *Find Your Power*.
- Adam Grant's *Give and Take*.
- For guidance on family boundary setting, see works by Nedra Glover Tawwab.
- For couples relationship skills, it's helpful to begin with the works of John and Julie Gottman.

Eco and Climate Action
One could stock an entire store only with books on sustainability and climate mitigation; there is an embarrassment of riches.

For realistically hopeful views of the big picture, see:
- Hannah Ritchie's *Not the End of the World*.
- Zahra Biabani's *Climate Optimism*.
- Susan Solomon's *Solvable* on the surprisingly successful human response to the ozone hole.
- Hans Rosling's *Factfulness* is a check on doomism. I especially appreciate his autobiographical *How I Learned to Understand the World*.
- With *I Want a Better Catastrophe*, Andrew Boyd demonstrates how to add some healthy humor to your grief and hope.
- For a good take on geoengineering, see Oliver Morton's *The Planet Remade*.

- For a wide-ranging climate history, Sunil Amrith's *The Burning Earth*.
- As a primer on basic meteorology and climate changes, John D. Cox's *Weather for Dummies* is a great recommendation.
- One of my favorite climate science books takes the form of a graphic novel, Philippe Squarzoni's *Climate Changed: A Personal Journey through the Science* (original French title: *Saison Brune*).
- For up-to-date explanations of climate breakdown and tipping points: Johan Rockstrom and Owen Gaffney's *Breaking Boundaries*.
- And for action steps based on climate science, see John Perona's *Knowledge to Power: The Comprehensive Handbook for Climate Science and Advocacy*.

Parenting and Child Development
- Works by educator David Sobel.
- Works by researcher Louise Chawla.
- Guides like Arielle Cook-Shonkoff's *Raising Anti-Doomers*.
- Mary DeMocker's *The Parents' Guide to Climate Revolution*.
- Leslie Davenport's *All the Feelings Under the Sun*.

Ethics
- Writings of Peter Singer.
- Travis Rieder's *Catastrophe Ethics*.
- Maggie FitzGerald's *Care and the Pluriverse*.

Action and Advocacy
- Margaret Klein Salamon's *Facing the Climate Emergency*.
- Adrienne Maree Brown's *Emergent Strategy*.
- Sarah Jaquette Ray's *A Field Guide to Climate Anxiety*.
- Dana Fisher's *Saving Ourselves*.
- Nick Hayes's *The Book of Trespass* and *Wild Service*.

Politics and Managing the Common Good
- Learn about Elinor Ostrom, the first woman to receive a Nobel Prize in economics. See *The Uncommon Knowledge of Elinor Ostrom*.
- For the US context, I have found Michael Maccoby's research

on narcissistic leaders in business and government helpful in understanding current events. I highly recommend *Narcissistic Leaders: The Incredible Pros, the Inevitable Cons.*
- The pragmatic views of Richard Rorty's *What Can We Hope For?*
- Eric Liu and Nick Hanauer's *The Gardens of Democracy.*
- Jennifer Pahlka's *Recoding America.*
- In terms of public service, Ronald Heifetz's work on adaptive leadership, including Ronald Heifetz and Marty Linsky, *Leadership on the Line: Staying Alive Through the Dangers of Change.*

Societal Critique
- See Astra Taylor's writings on solidarity and precarity; my late colleague Tod Sloan's *Life Choices* and *Damaged Life*; and Byung-chul Han, *The Disappearance of Rituals.*
- Critiques can also show us a way forward, as with the policy work of legislators like Alexandria Ocasio-Cortez and Ed Markey, the insights of David Graber and Malcolm Harris, and the writing of Pete Davis, including *Dedicated*, on a counterculture of commitment.

Arts and Culture
- Look for the works of Melanie Challenger.
- Donna Haraway's *Staying with the Trouble.*
- *Climate Lyricism* by Min Hyoung Song.
- Consult Matthew Schneider-Mayerson and Brent Ryan Bellamy's *An Ecotopian Lexicon.*
- Alison Hawthorne Deming and Lauret Savoy's *The Colors of Nature.*
- Debra Rosenthal's *Teaching the Literature of Climate Change.*

Musical Works
- Scott Ordway's *The End of Rain.*

Film
- *Minari* by Lee Isaac Chung is a nuanced portrayal of a Korean immigrant family's acculturation to the US and connection to land and nature.

- *Safe* by Tod Haynes – a prescient film anticipating eco-anxiety and environmental sensitivities.
- For an uplifting story of family and community survival in Africa, see Chiwetel Ejiofor's *The Boy Who Harnessed the Wind*.

Memoir
- *We Will Be Jaguars* by Nemonte Nenquimo.

Novels
- Of domestic life: Emma Pattee's *Tilt*.
- For young readers: Alan Gratz's *Two Degrees*.
- Cautionary tales: *The Wall* by John Lanchester.
- As a guide: *The Ministry for the Future* by Kim Stanley Robinson.

Nature Poetry Beyond the White European Canon
- Camille Dungy's anthology *Black Nature: Four Centuries of African American Nature Poetry*.
- Joy Harjo's *When the Light of the World Was Subdued, Our Songs Came Through: A Norton Anthology of Native Nations Poetry*.

Earth-Centered Spirituality
- Pope Francis's encyclical *Laudato Sí: On Care for Our Common Home*.
- Joanna Macy's *World as Lover, World as Self*.
- Debra Riensra's *Refugia Faith*.
- Bron Taylor's *Dark Green Religion*.

A Final Thought
A teacher once told me: When you are doing research and can't find the next reference, or the next study, or the next program, or the next invention, that's a signal that you need to create it.

NOTES

Introduction

1. Theodore Roszak, *Person/Planet: The Creative Disintegration of Industrial Society* (Garden City, NY: Anchor Press, 1978), xix.

2. Thomas J. Doherty and Susan Clayton, "The Psychological Impacts of Global Climate Change," *American Psychologist* 66, no. 4 (2011): 265–276, https://doi.org/10.1037/a0023141.

3. David Wallace-Wells, *The Uninhabitable Earth: Life After Warming* (New York: Crown Books, 2019), 3.

4. Rachel Carlson, *The Sense of Wonder: A Celebration of Nature for Parents and Children* (New York: Harper Perennial, 2017), 44.

5. "L'objet de la psychologie est de nous donner une idée toute autre des choses que nous connaissons le mieux," Paul Valéry, *Tel Quel...*(Paris: Gallimard, 1944), 279.

6. Mary Hawk et al., "Harm Reduction Principles for Healthcare Settings," *Harm Reduction Journal* 14 (2017): 70, https://doi.org/10.1186/s12954-017-0196-4.

7. Roger A. Pielke, *The Honest Broker: Making Sense of Science in Policy and Politics* (Cambridge, UK: Cambridge University Press, 2007).

8. David Abram, *The Spell of the Sensuous: Perception and Language in a More-Than-Human World* (New York: Pantheon Books, 1996).

9. Henry David Thoreau, *Walden* (Boston: Beacon Press, 2016), 89.

Part One: Coping
Chapter 1: Thinking

1. Daniel Sherrell, *Warmth: Coming of Age at the End of Our World* (New York: Penguin Books, 2021), 246.

2. The concept of "solastalgia" was developed by Australian philosopher Glenn Albrecht after observing the distress of individuals in the Hunter Valley, one of the largest open pit coal mining areas in the world.

3. The concept of "scale vertigo" comes from a conversation with environmental writer Emma Marris, who has explored creative ways to perceive local natural scales in books like *Wild Souls: Freedom and Flourishing in the Non-Human World* (New York: Bloomsbury, 2021).

4. Timothy Morton, *The Ecological Thought* (Cambridge, MA: Harvard University Press, 2010), 19.

5. See the extensive work of social psychologist Carol Dweck on the role of mindset in well-being and performance, including *Mindset: The New Psychology of Success* (New York: Random House, 2008).

6. See Brett Jenks, "To Make Progress on Climate Action, Pop 'Normative Bubbles,'" *Behavioral Scientist*, January 30, 2023, https://behavioralscientist.org/to-make-progress-on -climate-action-pop-normative-bubbles.

7. E. L. Doctorow, "Interview," in *Writers at Work: The Paris Review Interviews, 2nd Series*, edited by George Plimpton (New York: Viking Press, 1963).

Chapter 2: Feeling

1. Associated with Steven Hayes, founder of ACT Therapy. See Hayes's TEDx Talk "Psychological Flexibility: How Love Turns Pain into Purpose," YouTube, posted by TEDx Talks, February 22, 2016, https://www.youtube.com/watch?v=o79_gmO5ppg.

2. Aldo Leopold, *A Sand County Almanac: With Essays on Conservation from Round River* (New York: Ballantine Books, 1966), 197.

3. An example of the "despair and empowerment" approach can be found in the work of Buddhist philosopher Joanna Macy. See Joanna Macy, *Despair and Empowerment in the Nuclear Age* (Philadelphia: New Society, 1983) and other works.

Chapter 3: Calming

1. "Wisdom Weavings," Tribal Trust Foundation, https://tribaltrustfoundation.org /safeguarding-indigenous-wisdom-in-bhutan, accessed April 21, 2025.

2. This definition was put forth by the 1987 United Nations Brundtland Commission. United Nations, "Sustainability," https://www.un.org/en/academic-impact/sustainability. See World Commission on Environment and Development, *Our Common Future* (Oxford: Oxford University Press, 1987), http://www.un-documents.net/our-common-future.pdf.

3. Robert Kegan, *In Over Our Heads: The Mental Demands of Modern Life* (Cambridge, MA: Harvard University Press, 1998).

4. Emotional flooding: See Elizabeth Earnshaw, "How Stress Affects Relationships," Gottman Institute, last updated August 30, 2024, https://www.gottman.com/blog/how -stress-can-cause-relationship-dissatisfaction.

5. L. H. Powell, "The Hook: A Metaphor for Gaining Control of Emotional Reactivity," in *Heart and Mind: The Practice of Cardiac Psychology*, edited by Robert Allan and Stephen S. Scheidt (Washington, DC: American Psychological Association, 1996).

6. Ashley McNeil, "Farming as a Political Act: The Connection Between African-Americans and Land," Corps Network, https://corpsnetwork.org/farming-as-a-political -act-the-connection-between-african-americans-and-land, accessed January 25, 2025.

7. Chellis Glendinning is the author of the classic societal critique *My Name Is Chellis and I'm in Recovery from Western Civilization* (Boston: Shambhala, 1994).

8. "Varieties of Hope with Guest Elin Kelsey," *Climate Change and Happiness* podcast, hosted by Thomas Doherty and Panu Pihkala, season 1, episode 16, August 5, 2022, https:// climatechangeandhappiness.com/episodes/episode-16-varieties-of-hope-with-guest-elin-kelsey.

Chapter 4: Adapting

1. Gary Snyder, *Turtle Island, with "Four Changes"* (New York: New Directions, 1974), 100.

2. For example, see Kate Whiting and HyoJin Park, "This Is Why 'Polycrisis' Is a Useful Way of Looking at the World Right Now," World Economic Forum, 2023, https://www.weforum .org/stories/2023/03/polycrisis-adam-tooze-historian-explains.

3. The Climate Vulnerability Index can be found at https://climatevulnerabilityindex.org.

4. The National Climate Assessment is at https://nca2023.globalchange.gov. See also Dan Pisut, "How to Use the National Climate Assessment Interactive Atlas Explorer," Esri, November 1, 2023, https://mediaspace.esri.com/media/t/1_m3iem21y.

5. Hop Hopkins, "Racism Is Killing the Planet," *Sierra: The Magazine of the Sierra Club*, June 8, 2020, https://www.sierraclub.org/sierra/racism-killing-planet.

6. See, for example, Bob Doppelt, *Transformational Resilience: How Building Human Resilience to Climate Disruption Can Safeguard Society and Increase Wellbeing* (Sheffield, UK: Greenleaf, 2016).

7. For a perspective, see Jem Bendell and Rupert J. Read, eds., *Deep Adaptation: Navigating the Realities of Climate Chaos* (Cambridge, UK: Polity, 2021).

Part Two: Identity
Chapter 5: Meaning

1. Margaret Klein Salamon with Molly Gage, *Facing the Climate Emergency: How to Transform Yourself with Climate Truth*, 2nd ed. (Gabriola Island, BC: New Society, 2023), 14.

2. For examples of Macy's work, see Joanna Macy and Molly Young Brown, *Coming Back to Life: The Updated Guide to the Work That Reconnects* (Gabriola Island, BC: New Society, 2014).

3. Background on this Buddhist concept: See Pierce Salguero, "What Is a Bodhisattva? A Scholar of Buddhism Explains," The Conversation, https://theconversation.com/what-is-a-bodhisattva-a-scholar-of-buddhism-explains-189366.

4. One of her inspirations was the Canadian nature painter Emily Carr. See Lisa Baldissera, *Emily Carr: Life and Work* (Toronto: Art Canada Institute/Institut de l'Art Canadien, 2021), also available at https://www.aci-iac.ca/art-books/emily-carr/biography.

Chapter 6: Values

1. W. B. Yeats, "Aedh Wishes for the Cloths of Heaven" (1899), in *The Wind Among the Reeds* (New York: Woodstock Books, 1994).

2. Obviously, the ecological paradigm described by sociologists was not actually new, as human kinship with nature is a perennial ideal and remains a basis of many cultures around the world.

For an overview of the research see: Riley E. Dunlap, "The New Environmental Paradigm Scale: From Marginality to Worldwide Use," *Journal of Environmental Education* 40, no. 1 (2008): 3–18, https://doi.org/10.3200/JOEE.40.1.3-18. For historical context on Oregon in the 1970s, see Thomas Doherty, "Riley Dunlap: The Ecopsychology Interview," *Ecopsychology* 3, no. 4 (2011): 219–226.

3. A norm is a standard for behavior that is considered typical or proper in a given context. For a current perspective, see Cecilia Heyes, "Rethinking Norm Psychology," *Perspectives on Psychological Science* 19, no. 1 (2024): 12–38, https://doi.org/10.1177/17456916221112075.

4. Incident and government response: Gareth Harris, "Protestors Who Poured Soup over Van Gogh's Sunflowers Sentenced to Prison," September 27, 2024, https://www.theartnewspaper.com/2024/09/27/just-stop-oil-activists-who-poured-soup-over-vincent-van-goghs-sunflowers-sentenced-to-prison. I actually see the soup-throwing as a form of iconoclastic art and believe Van Gogh would have as well.

5. For background, see P. Wesley Schultz, "The Structure of Environmental Concern: Concern for Self, Other People, and the Biosphere," *Journal of Environmental Psychology* 21 (2001): 1–13. For values applied to messaging: P. Wesley Schultz and Lynnette Zelezny, "Reframing Environmental Messages to Be Congruent with American Values," *Human Ecology Review* 10 (2003): 126–136.

6. In surfing, "the stoke" is a term for the good feelings or euphoria one feels while being on the water and catching waves. For a discussion, see Bron Taylor, "Surfing into Spirituality and a New, Aquatic Nature Religion," *Journal of the American Academy of Religion* 75, no.

4 (2007): 923–951, DOI: 10.1093/jaarel/lfm067. See also Thomas J. Doherty et al., "Surfing and Psychology: Surfers' Experience, Identity, and Empowerment in Context," panel at the annual meeting of the American Psychological Association, Honolulu, August 2013.

7. Information on the Master Recycler Program is at https://www.masterrecycler.org.

8. Services like Ridwell, available in some cities, add options for home recycling. David Riemer, "Ridwell Lands over 75,000 Recycling Customers by Meeting the Moment," *Forbes*, March 21, 2023, https://www.forbes.com/sites/davidriemer/2023/03/21/ridwell-lands-over -75000-recycling-customers-by-meeting-the-moment. (I myself use services like Ridwell.)

9. Adapted and expanded from Stephen R. Kellert, *The Value of Life: Biological Diversity and Human Society* (Washington, DC: Island Press, 1996), and Stephen R. Kellert, *Birthright: People and Nature in the Modern World* (New Haven, CT: Yale University Press, 2012).

10. For a study, see Matthew H. Goldberg et al., "Perceptions and Correspondence of Climate Change Beliefs and Behavior Among Romantic Couples," *Journal of Environmental Psychology* 82 (2022): 101836.

11. In these "hunting derbies," the "varmint" targets could include bobcats, foxes, raccoons, crows, prairie dogs, or wolves. These contests are legal in most of the United States, and many allow children to participate.

12. Václav Havel, *Disturbing the Peace: A Conversation with Karel Huizdala*, trans. Paul Wilson (New York: Vintage, 1991), 181.

Chapter 7: Nature

1. Ralph Ellison, *Invisible Man* (New York: Vintage Books, 1982), 564.

2. On Citizens' Climate Lobby, see the group's website at https://citizensclimatelobby .org. See also Kevin Ummel, "Household Impact Study II (HIS2): The Impact of a Carbon Fee and Dividend Policy on the Finances of U.S. Households," Citizens' Climate Lobby, August 2020, https://citizensclimatelobby.org/wp-content/uploads/2018/06/HIS2-Working-Paper -v1.1.pdf.

3. The term "political animal," meaning someone who is fascinated by politics and who thrives on being closely involved in politics, goes back to Aristotle (Taegan Goddard's Political Dictionary, https://politicaldictionary.com/words/political-animal, accessed April 19, 2025).

4. John Schwartz, "Katharine Hayhoe, a Climate Explainer Who Stays Above the Storm," *New York Times*, October 10, 2016, https://www.nytimes.com/2016/10/11/science /katharine-hayhoe-climate-change-science.html. The Climate Reality Project was formed in 2011. See the project's website at https://www.climaterealityproject.org/our-mission.

5. For a natural history of Walden Pond, see "Walden Pond, Massachusetts: Environmental Setting and Current Investigations," US Geological Survey, 1998, https://pubs.usgs .gov/fs/1998/fs064-98/pdf/fs06498.pdf.

6. In fact, there is a small lake in Ontario, Canada, not unlike Walden Pond, that is used as a global geological marker of the Anthropocene era. Evidence of major human-caused changes to Earth in the twentieth century, such as acid rain and fallout from faraway nuclear weapons testing, is preserved in precise year-by-year layers in its deep sediments. Conservation Halton, "Crawford Lake Studies," ttps://www.conservationhalton.ca/crawford-lake -studies, accessed April 19, 2025; Alanna Mitchell, "The Anthropocene Is Here—and Tiny Crawford Lake Has Been Chosen as the Global Ground Zero," *Canadian Geographic*, last updated March 26, 2024, https://canadiangeographic.ca/articles/the-anthropocene-is-here -and-tiny-crawford-lake-has-been-chosen-as-the-global-ground-zero.

7. On Ulrich's career, see Sara O. Marberry, "Roger Ulrich Discusses Evidence-Based Design in Healthcare," *Healthcare Design*, October 31, 2010, https://healthcaredesignmagazine .com/trends/architecture/conversation-roger-ulrich.

8. See Frances E. Kuo, William C. Sullivan, Rebekah Levine Coley, and Liesette Brunton, "Fertile Ground for Community: Inner-City Neighborhood Common Spaces," *American Journal of Community Psychology* 26 (1998): 833–851.

9. Peter H. Kahn et al., "A Plasma Display Window? The Shifting Baseline Problem in a Technologically Mediated Natural World," *Journal of Environmental Psychology* 28, no. 2 (2008): 192–199.

10. Stephen Kellert, *Nature by Design: The Practice of Biophilic Design* (New Haven, CT: Yale University Press, 2018).

11. Mathew P. White et al., "Spending at Least 120 Minutes a Week in Nature Is Associated with Good Health and Wellbeing," *Scientific Reports* 9, no. 1 (2019): 7730, DOI: 10.1038 /s41598-019-44097-3. See also Genevive R. Meredith et al., "Minimum Time Dose in Nature to Positively Impact the Mental Health of College-Aged Students, and How to Measure It: A Scoping Review," *Frontiers in Psychology* 10 (2019), https://www.frontiersin.org/journals /psychology/articles/10.3389/fpsyg.2019.02942.

12. Erica Berry, *Wolfish: Wolf, Self, and the Stories We Tell About Fear* (New York: Flatiron Books, 2023).

13. There is a long history of research on this topic. See, for example, Ellen Cole, Esther D. Rothblum, and Eve M. Tallman, *Wilderness Therapy for Women: The Power of Adventure* (New York: Routledge, 1994); Irene G. Powch, "Wilderness Therapy: What Makes It Empowering for Women?," *Women and Therapy* 15, nos. 3–4 (1994): 11–27.

14. See "Hiking Safety: Another Look at the 10 Essentials," American Hiking Society, https://americanhiking.org/blog/hiking-safety-10-essentials, accessed January 28, 2025.

15. Originating in Japan, forest bathing is a style of therapeutic engagement with nature and forests, focusing on meditation, sensory experiences like walking barefoot, and the healing aspect of chemicals associated with trees. For an introduction, see Yoshifumi Miyazaki, *Shinrin-Yoku: The Japanese Art of Forest Bathing* (Portland, OR: Timber Press, 2018).

16. Kathryn Williams and David Harvey, "Transcendent Experience in Forest Environments," *Journal of Environmental Psychology* 21, no. 3 (2001): 249–260, https://doi.org /10.1006/jevp.2001.0204.

17. In this incident, a Black man birdwatching in New York's Central Park was falsely accused by a white woman of threatening her. See Christian Cooper, *Better Living Through Birding: Notes from a Black Man in the Natural World* (New York: Random House, 2023).

18. From the perspective of the US Environmental Protection Agency, environmental justice means the just treatment and meaningful involvement of all people, regardless of income, race, color, national origin, tribal affiliation, or disability, in decision-making that affects human health and the environment, so that people are fully protected from disproportionate and adverse human health and environmental effects, including those related to climate change. See "Learn About Environmental Justice," United States Environmental Protection Agency, last updated November 5, 2024, https://www.epa.gov/environmentaljustice /learn-about-environmental-justice (at the time of writing, this article was no longer available due to censorship and / or defunding).

19. Majora Carter's iconic TED talk "Greening the Ghetto" (February 2006) raised consciousness about environmental justice for many people. See https://www.ted.com/talks/majora_carter_greening_the_ghetto.

20. "Wood Products Continue to Grow Oregon's Economy," Oregon Forests Forever, October 2021, https://oregonforestsforever.com/wood-products-continue-to-grow-oregons-economy; "Mass Timber Building and Manufacturing Tours," March 26, 2024, event associated with the 2024 International Mass Timber Conference, Portland, OR, https://masstimberconference.com/panel/mass-timber-building-manufacturing-tours.

21. A project on Interstate 90 in Washington state's Cascade Mountains was the subject of a January 2018 documentary, *Cascade Crossroads*, https://conservationnw.org/our-work/habitat/cascade-crossroads-film. See also the Pigeon River Gorge project near the Great Smoky Mountains: https://smokiessafepassage.org.

22. Olson's disorder was later diagnosed as Charcot-Marie-Tooth disease. Shelley Essak, "The Story Behind 'Christina's World' by Andrew Wyeth," ThoughtCo., last updated August 8, 2019, https://www.thoughtco.com/christinas-world-by-andrew-wyeth-183007. A novel inspired by Wyeth's painting and Olson's story is Christina Baker Kline, *A Piece of the World* (New York: William Morrow, 2017).

23. "Antaeus," *Encyclopaedia Britannica*, https://www.britannica.com/topic/Antaeus, accessed April 19, 2025. For a discussion of the Antaeus story in the context of climate coping, see Meera Subramanian, "Leap," in *The World as We Knew It: Dispatches from a Changing Climate*, edited by Amy Brady and Tajja Isen (New York: Catapult, 2022).

Chapter 8: Family

1. Richard Louv, *Last Child in the Woods: Saving Our Children from Nature-Deficit Disorder*, updated ed. (Chapel Hill, NC: Algonquin Books of Chapel Hill, 2008), 316.

2. This was when I had the opportunity to study with environmental educators like Mitchell and Cindy Thomashow. See Mitchell Thomashow, *Bringing the Biosphere Home* (Cambridge, MA: MIT Press, 2003).

3. Famous examples are the Swedish activist Greta Thunberg and, more recently, the Fijian activist Timoci Naulusala. On the latter, see "Timoci Naulusala," Pacific Cooperation Foundation, https://www.pcf.org.nz/pacific-climate-change-voices/timoci-naulusala, accessed January 28, 2025.

4. Yale and George Mason have been regularly tracking attitudes about climate change in the United States and other nations since 2008. A recent summary: Leiserowitz et al., "Climate Change in the American Mind: Beliefs & Attitudes, Fall 2024," Yale Program on Climate Change Communication and George Mason University Center for Climate Change Communication, 2024.

5. For examples of 9/11 trauma research among those exposed at a distance, such as through media, see Chatterjee et al. "Risk Factors for Depression Among Civilians After the 9/11 World Trade Center Terrorist Attacks: A Systematic Review and Meta-Analysis," *PLoS Currents*, March 30, 2018, DOI: 10.1371/currents.dis.6a00b40c8ace0a6a0017361d7577c50a.

6. The quote is from David Sobel, "Beyond Ecophobia: Reclaiming the Heart in Nature Education," *Orion*, Autumn 1995. This approach is illustrated in other works by Sobel, such as *Beyond Ecophobia: Reclaiming the Heart in Nature Education* (Great Barrington, MA: Orion Society, 1996) and *Children's Special Places: Exploring the Role of Forts, Dens, and Bush Houses in Middle Childhood* (Detroit: Wayne State University Press, 2002).

7. For an example of Louise Chawla's original research, see Louise Chawla, "Childhood Experiences Associated with Care for the Natural World: A Theoretical Framework

for Empirical Results," *Children, Youth and Environments* 17, no. 4 (2007): 144–170. See also "Children and Nature with Louise Chawla," *Climate Change and Happiness* podcast, hosted by Thomas Doherty and Panu Pihkala, season 2, episode 22, June 23, 2023, https://climate changeandhappiness.com/episodes/children-and-nature-with-louise-chawla.

8. For example, Charlotte A. Jones and Aidan Davison, "Disempowering Emotions: The Role of Educational Experiences in Social Responses to Climate Change," *Geoforum* 118 (2021): 190–200. See also Caroline Hickman et al., "Climate Anxiety in Children and Young People and Their Beliefs About Government Responses to Climate Change: A Global Survey," *The Lancet Planetary Health* 5, no. 12 (2021): e863–e873, DOI: 10.1016/S2542-5196(21)00278-3.

9. I learned of the "crossing paths" image from Laurence Steinberg with Wendy Steinberg, *Crossing Paths: How Your Child's Adolescence Triggers Your Own Crisis* (New York: Simon & Schuster, 1994).

10. "Climate Change, Children and a Better World with Guest Dr. Jade Sasser" and "Finding Meaning in 'Generation Dread' with Guest Britt Wray," *Climate Change and Happiness* Podcast, hosted by Thomas Doherty and Panu Pihkala, season 1, episode 8, April 15, 2022, https://climatechangeandhappiness.com/episodes/episode-08-climate-change-children -and-a-better-world-dr-jade-sasser, and season 1, episode 9, April 29, 2022, https://climate changeandhappiness.com/episodes/episode-09-finding-meaning-in-generation-dread -britt-wray.

Part Three: Healing

1. Ecotherapy brings nature into the counseling and psychotherapy process, as with outdoor or walking therapy, and uses therapy techniques to promote healthy connection with nature and address environmental concerns, as with climate therapy. See Thomas J. Doherty, "Theoretical and Empirical Foundations for Ecotherapy," in *Ecotherapy: Theory, Research and Practice*, edited by Martin Jordan and Joe Hinds (London: Palgrave, 2016).

Chapter 9: Hostage

1. Omar El Akkad (@omarelakkad), "One day, when it's safe, when there's no personal downside to calling a thing what it is, when it's too late to hold anyone accountable, everyone will have always been against this," X, October 5, 2023, https://x.com/omarelakkad/status /1717082321445421056.

2. Delia O'Hara, "Thomas Doherty Works at the Intersection of Psychology and Environmental Science," American Psychological Association, 2018, https://www.apa.org/members /content/doherty-psychology-environmental-science.

3. "The personal is political" is the insight that many personal experiences (particularly those of women) can be traced to one's location within a system of power relationships. The concept was first popularized in 1970 by Carol Hanisch; see her "The Personal Is Political," January 2006, https://webhome.cs.uvic.ca/~mserra/AttachedFiles/PersonalPolitical.pdf.

4. There is both strong consensus (agreement) and consilience (convergence of multiple kinds of evidence, from tree rings to ice cores to CO_2) regarding the basic science of human-caused global heating and climate breakdown. For an explanation, see Michael Shermer, "Consilience and Consensus" (2015), https://michaelshermer.com/sciam-columns/consilience -and-consensus.

5. Geoffrey Supran and Naomi Oreskes, "Assessing ExxonMobil's Climate Change Communications (1977–2014)," *Environmental Research Letters* 12 (2017): 084019. Many scientists who helped create climate science were employed by oil and gas companies: Ben Franta, "Early Oil Industry Knowledge of CO_2 and Global Warming," *Nature Climate Change* 8 (2018): 1024–1025.

6. On ancient observatories, see Diana Hubbell, "9 Observatories Where Ancient Humans Looked to the Stars," *Atlas Obscura*, https://www.atlasobscura.com/lists/ancient -observatories, accessed February 10, 2025.

7. Ecological rights ensured by governments: see Yann Aguil, "The Right to a Healthy Environment," International Union for Conservation of Nature, 2021, https://iucn.org/news /world-commission-environmental-law/202110/right-a-healthy-environment.

8. Romany Webb, "Environmental Rights in State Constitutions," *Climate Law: A Sabin Center Blog*, Columbia Law School, August 31, 2021, https://blogs.law.columbia .edu/climatechange/2021/08/31/environmental-rights-in-state-constitutions.

9. Amy Beth Hanson, "Montana Supreme Court Upholds State Judge's Landmark Ruling in Youth Climate Case," *PBS NewsHour*, December 18, 2024, https://www.pbs.org/newshour /nation/montana-supreme-court-upholds-state-judges-landmark-ruling-in-youth-climate -case#.

10. Sally Weintrobe, "The Culture of Uncare" (blog post), November 29, 2014, https:// www.sallyweintrobe.com/29-nov-2014-the-culture-of-uncare-bob-gosling-memorial -lecture; "On the Psychological Roots of the Climate Crisis with Sally Weintrobe," *Climate Change and Happiness* podcast, hosted by Thomas Doherty and Panu Pihkala, season 4, episode 2, September 13, 2024, https://climatechangeandhappiness.com/episodes/season-4 -episode-2-on-the-psychological-roots-of-the-climate-crisis-with-sally-weintrobe.

11. Riley E. Dunlap, Aaron M. McCright, and Jerrod H. Yarosh, "The Political Divide on Climate Change: Partisan Polarization Widens in the U.S.," *Environment: Science and Policy for Sustainable Development* 58, no. 5 (2016): 4–23, https://doi.org/10.1080/00139157.2016 .1208995.

12. S. Fred Singer: SourceWatch, https://www.sourcewatch.org/index.php/S._Fred_Singer, accessed February 10, 2024. Recent Heartland Institute efforts to distort climate science: Scott Waldman, "Climate Denial Group Wants to Subvert NOAA Data with Its Own," *E&E News*, May 31, 2024, https://www.eenews.net/articles/climate-denial-group-wants-to -subvert-noaa-data-with-its-own.

13. The *Drilled* podcast by journalist Amy Westerveldt is a particularly helpful resource that uncovers the history of industry and government-abetted climate change propaganda. Orient with season 1: https://drilled.media/podcasts/drilled.

14. On precarity, see Astra Taylor, *The Age of Insecurity: Coming Together as Things Fall Apart* (Toronto: House of Anansi Press, 2023).

15. In 2002, Greenpeace awarded BP CEO Lord Browne an Earth Day "Oscar" for Best Impression of an Environmentalist. Joe Nocera, "Green Logo, but BP Is Old Oil," *New York Times*, August 12, 2006, https://www.nytimes.com/2006/08/12/business/worldbusiness /green-logo-but-bp-is-old-oil.html.

16. Dave Foreman, "Earth First Statement of Principles and Membership Brochure," September 1980, Rachel Carson Center, Ludwig-Maximilians-Universität, Munich.

17. Climate science and teaching standards: Christopher Thomas Holland, "The Implementation of the Next Generation Science Standards and the Tumultuous Fight to Implement Climate Change Awareness in Science Curricula," *Brock Education Journal* 29, no. 1 (2020): 35, DOI: 10.26522/brocked.v29i1.646.

18. American Public Health Association, "A Public Health Approach to Building Mental Wellness and Resilience in the Face of the Climate Crisis: Recommendations for Community Groups," 2024, https://www.apha.org/-/media/files/pdf/topics/climate/cche_mental_health _product.pdf.

19. Matthew Olay, "DOD Combines Adaptation, Mitigation to Confront Climate

Change," Department of Defense, April 23, 2024, https://www.defense.gov/News/News-Stories /Article/Article/3753400/dod-combines-adaptation-mitigation-to-confront-climate -change; Sherri Goodman, *Threat Multiplier: Climate, Military Leadership, and the Fight for Global Security* (Washington, DC: Island Press, 2024).

20. You can complete the Six Americas Survey questions here: Breanne Chryst et al., "Six Americas Super Short Survey (SASSY!)," Yale Program on Climate Change Communication, https://climatecommunication.yale.edu/visualizations-data/sassy, accessed February 10, 2025.

21. David Graeber, *The Utopia of Rules: On Technology, Stupidity, and the Secret Joys of Bureaucracy* (Brooklyn, NY: Melville House, 2015), 89.

Chapter 10: Anxiety

1. Shierry Weber Nicholson, *The Love of Nature at the End of the World: The Unspoken Dimensions of Environmental Concern* (Cambridge, MA: MIT Press, 2003), 10.

2. Dan Kahan, "Why We Are Poles Apart on Climate Change," *Nature* 488, no. 7411 (2012): 255, https://doi.org/10.1038/488255a.

3. Daniel Sherrell, *Warmth: Coming of Age at the End of Our World* (New York: Penguin Books, 2021), 248.

4. Neuroscientist Lisa Feldman Barrett makes this point on the podcast *ReThinking with Adam Grant*, January 17, 2024, https://www.ted.com/podcasts/rethinking-with-adam-grant /you-have-more-control-over-your-emotions-lisa-feldman-barrett-transcript. She notes that recategorizing anxious arousal as something besides anxiety, such as determination, allows you to create a different reality and act differently.

5. A good general reference for exposure therapy is Jonathan S. Abramowitz, Brett J. Deacon, and Stephen P. H. Whiteside, *Exposure Therapy for Anxiety: Principles and Practice*, 2nd ed. (New York: Guilford Press, 2019).

6. Poet Rainer Maria Rilke wrote to nineteen-year-old Franz Xaver Kappus, "I want to beg you, as much as I can, dear sir, to be patient toward all that is unsolved in your heart and to try to *love the questions themselves* like locked rooms and like books that are written in a very foreign tongue. Do not now seek the answers, which cannot be given you because you would not be able to live them. And the point is, to live everything. *Live* the questions now. Perhaps you will then gradually, without noticing it, live along some distant day into the answer." Rainer Maria Rilke, *Letters to a Young Poet*, revised ed., trans. M. D. Herder (New York: W. W. Norton, 2004), 27.

7. Cristina Zarbo, Giorgio A. Tasca, Francesco Cattafi, and Angelo Compare, "Integrative Psychotherapy Works," *Frontiers in Psychology* 6 (2016): 2021, https://doi.org/10.3389 /fpsyg.2015.02021.

8. Russ Harris, *ACT Made Simple: An Easy-to-Read Primer on Acceptance and Commitment Therapy*, 2nd ed. (Oakland, CA: New Harbinger, 2019).

9. "'CHASING ICE' Captures Largest Glacier Calving Ever Filmed—OFFICIAL VIDEO," YouTube, posted by Exposure Labs, December 14, 2012, https://www.youtube .com/watch?v=hC3VTgIPoGU.

10. Sophie Olivia Hanson, "How Self-Compassion Can Help Activists Avoid Burnout," *Greater Good Magazine*, May 10, 2024, https://greatergood.berkeley.edu/article/item/how_self _compassion_can_help_activists_avoid_burnout.

11. Kate F. Hays, ed., *Performance Psychology in Action: A Casebook for Working with Athletes, Performing Artists, Business Leaders, and Professionals in High-Risk Occupations* (Washington, DC: American Psychological Association, 2009), https://doi.org/10.1037/11876-000.

Chapter 11: Despair

1. Daniel Sherrell, *Warmth: Coming of Age at the End of Our World* (New York: Penguin Books, 2021), 26. In discussing "the one best prepared for the apocalypse," Sherrell was referencing the character Justine in Lars Van Trier's 2011 climate allegory film *Melancholia*, https://www.imdb.com/title/tt1527186.

2. Sherrell, *Warmth*, 246.

3. Jenny Gross, "Beloved Tree in England Is Felled in 'Act of Vandalism,'" *New York Times*, September 28, 2023, https://www.nytimes.com/2023/09/28/world/europe/sycamore-gap-tree-uk.html; Amanda Urich, "Devastated LA Residents See Outpouring of Support: 'One of the More Beautiful Things I've Seen,'" *The Guardian*, January 19, 2025, https://www.theguardian.com/us-news/2025/jan/19/la-fire-residents-support-network.

4. Disenfranchised grief is associated with losses that are not openly acknowledged or mourning processes after a loss that are not recognized or supported (such as after miscarriage, loss of companion animals, etc.). See Gizem Cesur-Soysal and Ela Arı, "How We Disenfranchise Grief for Self and Other: An Empirical Study," *OMEGA—Journal of Death and Dying*, 89, no. 2 (2024): 530–549, https://doi.org/10.1177/00302228221075203. In a climate context: Panu Pihkala, "Climate Anxiety, Maturational Loss, and Adversarial Growth," *Psychoanalytic Study of the Child* 77, no. 1 (2024): 369–388, https://doi.org/10.1080/00797308.2023.2287382.

5. Charlene Luchterhand, "Grief Reactions, Duration, and Tasks of Mourning," US Department of Veterans Affairs, 2019, https://www.va.gov/WHOLEHEALTHLIBRARY/tools/grief-reactions-duration-and-tasks-of-mourning.asp.

6. Panu Pihkala, "Ecological Sorrow: Types of Grief and Loss in Ecological Grief," *Sustainability* 16, no. 2 (2024): 849, https://doi.org/10.3390/su16020849.

7. Laura A. King and Joshua A. Hicks, "Whatever Happened to 'What Might Have Been'? Regrets, Happiness, and Maturity," *American Psychologist* 62, no. 7 (2007): 625–636, https://existentialpsych.sites.tamu.edu/wp-content/uploads/sites/152/2016/08/KingHicks2007AP.pdf.

8. Jane Goodall and Douglas Abrams with Gail Hudson, *The Book of Hope: A Survival Guide for Trying Times* (New York: Celadon, 2021), xii.

9. Greta Thunberg, "School Strike for Climate: Save the World by Changing the Rules," TED Talks, December 12, 2019, transcript at https://www.rev.com/transcripts/greta-thunberg-ted-talk-transcript-school-strike-for-climate.

10. Dick Roy, "Maintaining Hope," *Oregon State Bar Bulletin*, August/September 2017, 70.

11. The We'Moon website is https://wemoon.ws.

12. Carrie Arnold, "Want to Reset Your Circadian Rhythms? Go Camping," Sleep.com, June 29, 2022, https://www.sleep.com/travel/good-sleep-camping.

13. Ed Simon, "What Viktor Frankl's Logotherapy Can Offer in the Anthropocene," Aeon, February 11, 2020, https://aeon.co/ideas/what-viktor-frankls-logotherapy-can-offer-in-the-anthropocene.

14. Magdalena Budziszewska and Sofia Elisabet Jonsson, "From Climate Anxiety to Climate Action: An Existential Perspective on Climate Change Concerns Within Psychotherapy," *Journal of Humanistic Psychology*, published online February 10, 2021, https://doi.org/10.1177/0022167821993243.

15. See, for example, Mark Williams, John Teasdale, Zindel Segal, and Jon Kabat-Zinn, *The Mindful Way Through Depression: Freeing Yourself from Chronic Unhappiness* (New York: Guilford Press, 2007).

16. Marc O. Williams and Victoria M. Samuel, "Acceptance and Commitment Therapy as an Approach for Working with Climate Distress," *Cognitive Behaviour Therapist* 17 (2024): e35, DOI: 10.1017/S1754470X23000247. The image of "dropping the rope" is associated with ACT theorist Steven Hayes.

Chapter 12: Place

1. Melanie Challenger, *How to Be Animal: A New History of What It Means to Be Human* (New York: Penguin Books, 2021), 1.

2. See David W. Orr, *On Education, Environment, and the Human Prospect* (Washington, DC: Island Press, 2004).

3. I first became aware of this cycle after reading the work of David Kidner. See David W. Kidner, *Environmentalism and the Politics of Subjectivity* (Albany: State University of New York Press, 2001).

4. D. Y. Jayakody, V. M. Adams, G. Pecl, and E. Lester, "What Makes a Place Special? Understanding Drivers and the Nature of Place Attachment," *Applied Geography* 163 (2024): 103177, https://doi.org/10.1016/j.apgeog.2023.103177.

5. "Our Emotional Attachment to Nature with Susan Bodnar," *Climate Change and Happiness* podcast, hosted by Thomas Doherty and Panu Pihkala, season 2, episode 16, March 31, 2023, https://climatechangeandhappiness.com/episodes/season-2-episode-16-our-emotional -attachment-to-Nature-with-susan-bodnar. Said Bodnar: "We started with the simplest of questions. 'Think of a place…What does it remind you of?' People said, 'Mother, father, mentor, best friend, sibling'…'If this place were no longer here, how would you feel?' 'Devastated'…'What else devastates you?'…The loss of someone you love."

6. On the pilgrimage route, see "La Via di Francesco," https://www.viadifrancesco.it /en, accessed February 11, 2025; and Elle Bieling, "The Way of St. Francis—La Via di Francesco, Introduction," PilgrimageTraveler.com, https://www.pilgrimagetraveler.com/way-of-st -francis.html, accessed February 11, 2025.

7. Andy Fisher, *Radical Ecopsychology: Psychology in the Service of Life*, 2nd ed. (Albany: State University of New York Press, 2013), 13.

8. David Abram, *The Spell of the Sensuous* (New York: Vintage, 1997), 202.

9. Radical Joy for Hard Times, https://radicaljoy.org. See also "Radical Joy in the Midst of Environmental Grief with Trebbe Johnson," *Climate Change and Happiness* podcast, hosted by Thomas Doherty and Panu Pihkala, season 3, episode 15, March 16, 2024, https:// climatechangeandhappiness.com/episodes/season-3-episode-15-radical-joy-for-hard -times-with-trebbe-johnson.

10. A standard tribal land acknowledgment in the United States should include an account of local Native tribes and bands and the historical context. For example, while there are nine federally recognized tribes in the state of Oregon, many are "consolidated tribes" made up of remnants and refugees after the treaties of the 1850s took away Native land. The Portland metro area rests on traditional village sites of the Multnomah, Wasco, Cowlitz, Kathlamet, Clackamas, Bands of Chinook, Tualatin, Kalapuya, Molalla, and other peoples who made their homes along the Columbia River.

11. For a Maori person, their *turangawaewae* or standing place would usually be a traditional *marae*, a tribal gathering place on ancestral land. "Māori Manners and Social Behaviour— Ngā Mahi Tika," in *Te Ara—The Encyclopedia of New Zealand*, https://teara.govt.nz/en/maori -manners-and-social-behaviour-nga-mahi-tika/page-2, accessed April 21, 2025.

12. "Empowerment Economics and Portland's Native Community," NAYA Family Center, November 29, 2021, https://nayapdx.org/blog/2021/11/29/empowerment-economics -and-portlands-native-community.

13. "Return to Neerchokikoo," NAYA Family Center, https://nayapdx.org/support-us/return -to-neerchokikoo, accessed February 11, 2025.

14. Food sovereignty groups: Black to the Land Coalition, https://www.blacktothe landcoalition.com, and the National Black Farmers Association, https://www.blackfarmers .org. Initiatives specific to the Portland Oregon area include Black Futures Community Farm,

https://blackfutures.farm; Oregon State University's Black Food Sovereignty Coalition, https://smallfarms.oregonstate.edu/smallfarms/black-and-brown-farmers-color-oregon; and Black Farmers Portland's Come Through Market initiative, https://www.comethrupdx.org.

15. Section 22007 of the Inflation Reduction Act. "Discrimination Financial Assistance Program," USDA, https://www.usda.gov/dfap-foia, accessed April 20, 2025.

Part Four: Flourishing
Chapter 13: Happiness

1. Panu Pihkala and Thomas Doherty, "Eco-Anxiety and Happiness," in *Encyclopedia of Happiness, Quality of Life and Subjective Wellbeing*, edited by Hilke Brockmann and Roger Fernandez-Urbano (Cheltenham, UK: Edward Elgar, 2024).

2. Examples of parents for climate groups in the United States and worldwide: Our Kids' Climate, https://ourkidsclimate.org; Families for Climate, https://www.familiesforclimate.org; Climate Action Families, https://climateactionfamilies.org.

3. Jon Kabat-Zinn, *Full Catastrophe Living*, rev. ed. (New York: Bantam, 2013). The phrase "full catastrophe living" comes from the 1961 Michael Cacoyannis film *Zorba the Greek*: "Am I not a man? And is not a man stupid? I'm a man, so I'm married. Wife, children, house—everything. The full catastrophe." See "'The Full Catastrophe'—Zorba the Greek (clip)," YouTube, posted by Mindfulness 360—Center for Meditation, January 25, 2017, https://www.youtube.com/watch?v=x9Dy_5zMfEM.

4. The PERMA model identifies five factors that contribute to a sense of flourishing: positive emotions, engagement, relationships, meaning, and accomplishments. See Martin Seligman, *Flourish* (New York: Free Press, 2011).

5. The IPAT equation ($I = P \times A \times T$) maintains that impacts on ecosystems (I) are the product of the population size (P), affluence (A), and technology (T) of the human population in question. For an accessible explanation see "Population, Affluence, and Technology," part of the course materials for GEOG 30N, "Environment and Society in a Changing World," Penn State University, https://www.e-education.psu.edu/geog30/node/328#, accessed February 12, 2025.

6. Barbara Ehrenreich, *Bright-Sided: How the Relentless Promotion of Positive Thinking Has Undermined America* (New York: Metropolitan Books, 2009).

7. Friedrich Nietzsche, *Beyond Good and Evil: Prelude to a Philosophy of the Future*, edited by Rolf-Peter Horstmann and Judith Norman, trans. Judith Norman (Cambridge, UK: Cambridge University Press, 2002), chap. IV, sec. 146.

8. An example of wisdom research: Kaili Zhang, Juan Shi, Fengyan Wang, and Michel Ferrari, "Wisdom: Meaning, Structure, Types, Arguments, and Future Concerns," *Current Psychology* 42 (2023): 15030–15051.

9. Lorna Collier, "Growth After Trauma: Why Are Some People More Resilient Than Others—and Can It Be Taught?," *Monitor on Psychology* 47, no. 10 (2016): 48, https://www.apa.org/monitor/2016/11/growth-trauma.

10. See Beyza Sümer, "Ibn Khaldun's Asabiyya for Social Cohesion," *Elektronik Sosyal Bilimler Dergisi* 11, no. 41 (2012): 253–267, https://dergipark.org.tr/en/download/article-file/70392#.

11. On fire breathing (Sanskrit *kapalabhati*), see Rolf Sovik, "Learn Kapalabhati (Skull Shining Breath)," Yoga International, https://yogainternational.com/article/view/learn-kapalabhati-skull-shining-breath, accessed February 12, 2025.

12. Gurucharan Singh Khalsa and Yogi Bhajan, *Breathwalk: Breathing Your Way to a Revitalized Body, Mind, and Spirit* (New York: Crown, 2000).

13. Edward Abbey, "One Final Paragraph of Advice…," quoted in Reed F. Noss and Allen Y. Cooperrider, *Saving Nature's Legacy: Protecting and Restoring Biodiversity* (Washington, DC: Island Press, 1994), 338.

Chapter 14: Relationships

1. Quoted in Elisabeth Young-Bruehl, *Anna Freud: A Biography*, 2nd ed. (New Haven, CT: Yale University Press, 2008), 18.

2. Sociologist Kari Norgaard has documented the process of socially organized denial regarding climate change in Norway, a wealthy country that produces nearly 2 million barrels of oil a day from its extensive North Sea reserves. See Kari Norgaard, *Living in Denial: Climate Change, Emotions, and Everyday Life* (Cambridge, MA: MIT Press, 2011).

3. Bill Doherty cofounded the nonprofit Braver Angels. As he puts it, their goal "is not to change people's views of issues, but to change their views of each other": https://braverangels.org/our-mission. (He and the author are not related.)

4. Renée Lertzman, "The Myth of Apathy," *The Ecologist*, June 19, 2008, https://theecologist.org/2008/jun/19/myth-apathy. See also "Revisiting the Myth of Climate Apathy with Renée Lertzman," *Climate Change and Happiness* podcast, hosted by Thomas Doherty and Panu Pihkala, season 2, episode 24, July 21, 2023, https://climatechangeandhappiness.com/episodes/season-2-episode-24.

5. The original version addressed the fears of speaking out about nuclear war. See Joanna Macy, *Despair and Personal Power in the Nuclear Age* (Philadelphia: New Society, 1983).

6. Stephen R. Covey, *The 7 Habits of Highly Effective People* (New York: Fireside, 1989), 239.

7. The "tend and befriend" impulse is visible in the genuine acts of goodwill witnessed in every disaster. See Shelley E. Taylor, "Tend and Befriend," in *Handbook of Theories of Social Psychology, Volume 1*, edited by Paul A. M. Van Lange, Arie W. Kruglanski, and E. Tory Higgins (London: Sage, 2012), 32–49.

8. Yale University's School of the Environment offers programs in both forestry and industrial ecology: https://environment.yale.edu/learning-communities/forestry#related-centers. See also Oregon Forest Industry Directory, https://www.orforestdirectory.com/.

9. Oregon State University, College of Forestry, Wood Science and Engineering, https://wse.forestry.oregonstate.edu/graduate-programs.

10. I got a chance to hear Frank Luntz speak at the 2023 Aspen Ideas Conference. You can see it here: "Americans and Climate: Words to Use, Words to Lose," YouTube, posted by The Aspen Institute, June 27, 2022, https://www.youtube.com/watch?v=Kzj-Co0wy0s.

11. See Braver Angels' instructional videos "Families and Politics" and "Skills for Bridging the Divide," with versions for Red and Blue voters: https://braverangels.org/what-we-do/take-an-ecourse.

12. W. B. Yeats, "Aedh Wishes for the Cloths of Heaven" (1899), in *The Wind Among the Reeds* (New York: Woodstock Books, 1994).

13. Adam M. Grant, *Give and Take: A Revolutionary Approach to Success* (New York: Viking, 2013).

Chapter 15: Art

1. Ezra Pound, "Canto LXXXI," *The Cantos of Ezra Pound* (New York: New Directions, 1986), https://www.poetryfoundation.org/poems/54320/canto-lxxxi.

2. John Luther Adams won a Pulitzer Prize for *Become Ocean*, part of the Become trilogy with *Become River* and the sublime *Become Desert*.

3. Images of the dolphin fresco, created around 1600–1450 BCE, can be seen at Vir Muze, https://virmuze.com/m/minoan-mycenaean/x/minoan-fresco, and the Archaeological Museum of Heraklion, https://www.heraklionmuseum.gr/en/exhibit/dolphin-fresco.

4. Kelly M., "Arte-Factual: Minoan Dolphin Fresco," Tomb Raider Horizons, October 7, 2013, https://tombraiderhorizons.com/2013/10/07/arte-factual-minoan-dolphin-fresco.

5. "Brother Warrior," YouTube, posted by Kate Wolf—Topic, February 15, 2021, https://www.youtube.com/watch?v=gj89NFO7LFU.

6. *Wild Kingdom*, Wikipedia, https://en.wikipedia.org/wiki/Wild_Kingdom, accessed February 13, 2025.

7. *America Outdoors with Baratunde Thurston*, https://www.pbs.org/show/america-outdoors-baratunde-thurston

8. Daniella Molnar, https://www.danielamolnar.com; "Art Gives Ecological Grief a Body," *Climate Change and Happiness* podcast, hosted by Thomas Doherty and Panu Pihkala, season 2, episode 12, February 3, 2023, https://climatechangeandhappiness.com/episodes/season-2-episode-12-art-gives-ecological-grief-a-body-with-daniela-molnar.

9. "About Art Therapy," American Art Therapy Association, 2022, https://arttherapy.org/about-art-therapy.

10. Daniel J. Levitin, *I Heard There Was a Secret Chord: Music as Medicine* (New York: W. W. Norton, 2024).

11. Mary Oliver, "Wild Geese," *Wild Geese: Selected Poems* (Highgreen, UK: Bloodaxe, 2004).

12. "The Postman: Quotes," IMDB, https://www.imdb.com/title/tt0110877/quotes, accessed February 13, 2024.

13. Dark Mountain Literary Movement and publications: https://dark-mountain.net/about/manifesto.

14. "On Nature, Poetry, and Creativity with Kim Stafford," *Climate Change and Happiness* podcast, hosted by Thomas Doherty and Panu Pihkala, season 2, episode 6, November 11, 2022, https://climatechangeandhappiness.com/episodes/season-2-episode-6-on-the-power-of-poetry-with-kim-stafford.

15. Patricia Y. Sanchez, "Reading Fiction Early in Life Is Associated with a More Complex Worldview, Study Finds," PsyPost, August 25, 2022, https://www.psypost.org/reading-fiction-early-in-life-is-associated-with-a-more-complex-worldview-study-finds.

16. Kim Stanley Robinson, *The Ministry for the Future* (New York: Orbit, 2021); Alan Gratz, *Two Degrees* (New York: Scholastic, 2022).

17. Jeff Orlowski, dir., *Chasing Ice* (Exposure Labs, 2012).

18. See "On Women, Fear and Nature with Erica Berry," *Climate Change and Happiness* podcast, hosted by Thomas Doherty and Panu Pihkala, season 2, episode 20, May 26, 2023, https://climatechangeandhappiness.com/episodes/season-2-episode-20-on-women-fear-and-nature-with-erica-berry.

19. Maya Lin, *What Is Missing*, https://www.whatismissing.org.

20. Levitin, *I Heard There Was a Secret Chord*, 3.

21. In the Oregon Symphony's 2024/25 season, "the harmony, rhythm, and power of Nature take center stage," https://www.orsymphony.org/concerts-tickets/the-Nature-of-music.

22. See the essays included in Moby's albums *Everything Is Wrong* (Mute, 1995), *Animal Rights* (Mute, 1996), and *Play* (Mute, 1992). See also Björk, *Biophilia* (One Little Indian, 2011), and her dialogues with naturalist David Attenborough: *When Björk Met Attenborough*, dir. Louise Hooper, 2013, available on YouTube, https://www.youtube.com/watch?v=c_jVvTW8Oco.

23. Ayana Elizabeth Johnson, *What If We Get It Right? Visions of Climate Futures* (New York: One World, 2024).

24. Olalekan B. Jeyifous's website is https://jeyifous.us. Works by Jeyifous used in Johnson's book *What If We Get It Right?* are visible at https://jeyifous.us/PIONEER.

25. See Mary Mattingly, "Swale: A Floating Food Forest | BK Stories," YouTube, posted by BRIC TV, July 3, 2018, https://www.youtube.com/watch?v=USm73rzFt8c; see also Zoe

Lescaze, "How Should Art Reckon with Climate Change?," *New York Times*, March 25, 2022, https://www.nytimes.com/2022/03/25/t-magazine/art-climate-change.html.

26. Erica Berry, "Why We Need More Climate Change Love Stories," *Outside*, October 8, 2021, https://www.outsideonline.com/culture/essays-culture/climate-change-love-stories -fiction-books. See also Sueellen Campbell, "The Power of Love in the Fight Against Climate Change," Yale Climate Connections, February 14, 2023, https://yaleclimateconnections .org/2023/02/the-power-of-love-in-the-fight-against-climate-change.

27. Joan Walsh Anglund, *A Cup of Sun: A Book of Poems* (New York: Harcourt, Brace & World, 1967).

28. "Black Sabbath: 05. Age of Reason (13 Album)," YouTube, posted by Max Rockatansky, June 14, 2013, https://www.youtube.com/watch?v=UN2JETFKYaE.

29. "The Weather Station (Tamara Lindeman)—Robber (Official Video)," YouTube, posted by The Weather Station, October 13, 2020, https://www.youtube.com/watch?v=O J9SYLVaIUI.

30. "The Thermals—Here's Your Future @ Academy Dublin 2009," YouTube, posted by Conor, December 18, 2011, https://www.youtube.com/watch?v=4F1nR22w7w4.

31. "Time of No Reply," YouTube, posted by Nick Drake, July 30, 2018, https://youtu.be /-Gqci8bFYXk.

32. "There'll Be Some Changes Made," YouTube, posted by Mildred Bailey—Topic, November 8, 2014, https://www.youtube.com/watch?v=5eCXAX8kTTE.

33. "Joon 'Cruel Summer' (Official Video)," YouTube, posted by Italians Do It Better Music, November 18, 2020, https://www.youtube.com/watch?v=WJa3IyR8Tos.

34. "Music," What If We Get It Right? website, https://www.getitright.Earth/music-art, accessed February 13, 2025.

35. Suzanne Brooker, *The Elements of Landscape Oil Painting: Techniques for Rendering Sky, Terrain, Trees, and Water* (New York: Watson-Guptill, 2015), 1.

36. "Expressive Writing Can Help Your Mental Health, with James Pennebaker, PhD," *Speaking of Psychology* podcast, episode 277, March 2024, https://www.apa.org/news/podcasts /speaking-of-psychology/expressive-writing.

37. Julia Cameron, *The Artist's Way: A Spiritual Path to Higher Creativity* (Los Angeles: Jeremy P. Tarcher / Perigee, 1992).

38. Andy Goldsworthy's website is https://andygoldsworthystudio.com.

39. Anthropologist Amanda Stronza's photographic memorials are visible on her website at https://www.amandastronza.com/passions#memorials. See also the exhibition *Kindred Spirits*, featuring works by Stronza, Rachel Ivanyi, and Hannah Salyer, at https:// rachelivanyi.com/kindred-spirits.

Chapter 16: Spirit

1. Pope Francis, "Encyclical Letter *Laudato Sí* of the Holy Father Francis on Care for Our Common Home," May 24, 2015, para. 2, http://www.vatican.va/content/francesco/en /encyclicals/documents/papa-francesco_20150524_enciclica-laudato-si.html.

2. Katy Z. Allen, "A Call for a New Kind of Chaplain," EcoFaith Recovery, March 19, 2015, https://www.ecofaithrecovery.org/a-call-for-a-new-kind-of-chaplain-by-rabbi-katy -allen; Judith Cowles, "At the Bedside of Mother Earth: A Call to Eco-Chaplaincy," The Chaplaincy Institute, July 2008, https://chaplaincyinstitute.org/portfolio-items/at-the-bed side-of-mother-Earth-a-call-to-eco-chaplaincy.

3. See a fictionalized encounter like this in *Extrapolations*, the Netflix climate series, season 1, episode 3. "Why is God doing this to us?" asks Alana Goldblatt (Neska Rose) as a fifth question during a Passover seder scene; https://www.imdb.com/title/tt15721382/?ref_=ttep_ep_3.

4. For an example, see Adelle M. Banks, "COP28 to Have First-Ever 'Faith Pavilion' at a UN Climate Summit," November 13, 2023, Earthbeat: A Project of National Catholic Reporter, https://www.ncronline.org/earthbeat/faith/cop28-have-first-ever-faith-pavilion-un -climate-summit.

5. Patricia O'Connell Killen and Mark Silk, *Religion and Public Life in the Pacific Northwest: The None Zone* (Walnut Creek, CA: AltaMira, 2004); Kristin Joyner, ed., *Filling the Void: Voices from the None Zone* (Knoxville, TN: Market Square Books, 2019).

6. Michael Lipka and Claire Gecewicz, "More Americans Now Say They're Spiritual but Not Religious," Pew Research Center, September 6, 2017, https://www.pewresearch.org /short-reads/2017/09/06/more-americans-now-say-theyre-spiritual-but-not-religious.

7. Gallup, "Religion," https://news.gallup.com/poll/1690/religion.aspx, accessed February 14, 2025.

8. For example, Panu Pihkala, "Eco-Anxiety and Pastoral Care: Theoretical Considerations and Practical Suggestions," *Religions* 13, no. 3 (2022): 192; Erin Reid, "The Role of Religious Literacy to Address Eco-Anxiety," Center for Civic Religious Literacy, https://ccrl-clrc .ca/eco-anxiety, accessed February 14, 2025.

9. "How to Cope with Climate Anxiety, with Thomas Doherty, PsyD, and Ashlee Cunsolo, PhD," *Speaking of Psychology* podcast, episode 138, April 2021, https://www.apa.org/news /podcasts/speaking-of-psychology/eco-anxiety.

10. St. Casimir's RC Church in Buffalo, New York, is a Byzantine-style church completed in 1929 and based on the mosque of St. Sophia in Istanbul, Turkey. See images of nave murals and stained glass at https://buffaloah.com/a/cable/160/10_dome/dome.html.

11. See Bron Taylor, *Dark Green Religion: Nature Spirituality and the Planetary Future* (Berkeley: University of California Press, 2009). On stewardship values, see Yasmeen Mahnaz Faruqi, "Islamic View of Nature and Values: Could These Be the Answer to Building Bridges Between Modern Science and Islamic Science," *International Education Journal* 8, no. 2 (2007): 461–469, https://files.eric.ed.gov/fulltext/EJ834281.pdf.

12. Pope Francis, "Encyclical Letter *Laudato Sí*," para. 2. The Laudato Sí Movement defines ecological conversion as the "transformation of hearts and minds toward greater love of God, each other, and creation" and "a process of acknowledging our contribution to the social and ecological crisis and acting in ways that nurture communion: healing and renewing our common home." Jonathan Bradon, "What Is an Ecological Conversion?," Laudato Sí Movement, June 24, 2021, https://laudatosimovement.org/news/what-is-an-ecological-conversion-en-news.

13. Robin Wall Kimmerer, *Braiding Sweetgrass: Indigenous Wisdom, Scientific Knowledge and the Teaching of Plants* (Minneapolis, MN: Milkweed Editions, 2013).

14. Melinda Storie and Joanne Vining, "From Oh to Aha: Characteristics and Types of Environmental Epiphany Experiences," *Human Ecology Review* 24, no. 1 (2018): 155–180.

15. Joanna Macy and John Seed developed the Council of All Beings ritual. See John Seed, Joanna Macy, Pat Fleming, and Arne Naess, *Thinking Like a Mountain: Towards a Council of All Beings* (Philadelphia: New Catalyst Books, 2007). Facilitation instructions are at https://workthatreconnects.org/resources/council-of-all-beings.

16. Psilocybin has been employed as an entheogen, or substance used for spiritual insight and growth, by many cultures. Though still illegal at the federal level at the time of writing, psilocybin therapy has been legal in Oregon since 2020, and in Colorado since 2022.

17. Emily Willow, "How Psychedelic Therapy May Help with Climate Change Anxiety," *Washington Post*, November 3, 2023, https://www.washingtonpost.com/wellness/2023/11/03 /psychedelics-therapy-climate-change-eco-anxiety.

18. Hannes Kettner, Sam Gandy, Eline C. H. M. Haijen, and Robin L. Carhart-Harris, "From Egoism to Ecoism: Psychedelics Increase Nature Relatedness in a State-Mediated and

Context-Dependent Manner," *International Journal of Environmental Research and Public Health* 16, no. 24 (2019): 5147.

19. "Why We Love Drugs: Michael Pollan on America's Broken—but Improving—Relationship with Drugs," *The Gray Area* podcast, hosted by Sean Illing, July 20, 2021, https://www.vox.com/vox-conversations-podcast/22526097/vox-conversations-michael-pollan-this-is-your-mind-on-plants.

20. See William B. Parsons, *The Enigma of the Oceanic Feeling: Revisioning the Psychoanalytic Theory of Mysticism* (New York: Oxford University Press, 1999).

21. Helping children with climate anxiety: Leslie Davenport and Irma Ruggiero, *What to Do When Climate Change Scares You* (Washington, DC: Magination Press, 2024); Chelsea Harper, Brook Irwin, and Penny Hood, *I Love You Forever and Always—The Storybook Project: Helping Mamas Talk to Their Kiddos About Breast Cancer* (n.p.: Jessica Santos, 2019), https://www.thestorybookproject.org.

Part Five: Action

1. For a short history of the decline of the bald eagle, protection efforts, and the success in getting the bald eagle off the endangered species list, see Avian Report, "From Endangered to Recovered: A Timeline of the Bald Eagle's Journey," https://avianreport.com/endangered-bald-eagle, accessed April 21, 2025.

Chapter 17: Duty

1. One of the classic writings on duty is Cicero, *De Officis* (Cambridge, MA: Harvard University Press, 1913). Cicero suggests that duties can come from four different sources: as a result of being a human, as a result of one's particular place in life (one's family, one's country, one's job), as a result of one's character, or as a result of one's own moral expectations for oneself.

2. Frederick Douglass, "West India Emancipation," speech at Canandaigua, New York, August 3, 1857, https://www.blackpast.org/african-american-history/1857-frederick-douglass-if-there-no-struggle-there-no-progress.

3. Bhagavad Gita 18:47, https://www.holy-bhagavad-gita.org/chapter/18/verse/47.

4. Technically, this is referred to as motivational interviewing, a counseling technique that reduces defensiveness and helps people clarify their goals and commitments for change. See Florian E. Klonek, Arnelie V. Güntner, Nale Lehmann-Willenbrock, and Simone Kauffeld, "Using Motivational Interviewing to Reduce Threats in Conversations About Environmental Behavior," *Frontiers in Psychology* 6 (2015): 1015, DOI: 10.3389/fpsyg.2015.01015.

5. This is known as values-beliefs-norms theory in environmental psychology. Paul C. Stern et al., "A Value-Belief-Norm Theory of Support for Social Movements: The Case of Environmentalism," *Human Ecology Review* 6, no. 2 (1999): 81–97.

6. In psychological terms, you've developed a personal norm, and this activates your behavior. See George L. W. Perry et al., "Evaluating the Role of Social Norms in Fostering Pro-Environmental Behaviors," *Frontiers in Environmental Science* 9 (2021), DOI: 10.3389/fenvs.2021.620125.

7. See Jonathan Krasner, "The World Is Broken, So Humans Must Repair It: The History and Evolution of Tikkun Olam," The Jewish Experience, Brandeis University, May 22, 2023, https://www.brandeis.edu/jewish-experience/history-culture/2023/may/tikkun-olam-history.html.

8. Citizens' Climate Lobby is a nonpartisan advocacy group focused on solving the problem of climate change by creating policies using a market-based approach to price carbon pollution from fossil fuels and end subsidies to fossil fuel companies. Volunteers build relationships with elected officials, the media, and their local community. See https://citizens climatelobby.org/about-ccl, accessed February 19, 2025.

9. For an example, see Brad M. Maguth, Andrea Tomer, and Avery Apanius, "The Despair to Empowerment Curricular Curve in Global Education: Lessons from a Seventh-Grade World History Classroom," *The Social Studies* 110, no. 1 (2019): 43–50.

10. Willett Kempton and Dorothy C. Holland, "Identity and Sustained Environmental Practice," in *Identity and the Natural Environment: The Psychological Significance of Nature*, ed. Susan Clayton and Susan Opotow (Cambridge, MA: MIT Press, 2003).

11. The environmental justice movement, for example, sprang from awareness of unfair and racially biased siting of hazardous waste sites. See Robert D. Bullard, *Dumping in Dixie: Race, Class, and Environmental Quality* (Boulder, CO: Westview Press, 1994).

Chapter 18: Challenges

1. Portia Nelson, *There's a Hole in My Sidewalk: The Romance of Self-Discovery* (New York: Atria Books, 2018), xi. I first heard this story when I was an intern at the University of Massachusetts Medical Center's mindfulness program.

2. This quote plays off John 8:32 ("And you will know the truth, and it will set you free") and has been attributed to William Garfield. David Foster Wallace offered a variation: "The truth will set you free. But not until it is finished with you" (*Infinite Jest* [New York: Little, Brown, 1996], 9). All ring true for coping and engaging with climate issues.

3. A question is how might compassionate force protocols be developed for healthcare settings being ethically applied to action to limit climate destructive behaviors? For background, see Constance E. George, "What Might a Good Compassionate Force Protocol Look Like?," *AMA Journal of Ethics* 23, no. 4 (April 2021): E326–334, https://journalofethics.ama-assn .org/article/what-might-good-compassionate-force-protocol-look/2021-04. See also "Compassionate Force," special issue, *AMA Journal of Ethics* 23, no. 4 (April 2021): E287–370, https://journalofethics.ama-assn.org/issue/compassionate-force.

4. Robert J. Brulle, "Networks of Opposition: A Structural Analysis of U.S. Climate Change Countermovement Coalitions 1989–2015," *Sociological Inquiry* 91 (2021): 603–624.

5. "Fact Sheet | Fossil Fuel Subsidies: A Closer Look at Tax Breaks and Societal Costs (2019)," Environmental and Energy Study Institute, July 29, 2019, https://www.eesi.org/papers /view/fact-sheet-fossil-fuel-subsidies-a-closer-look-at-tax-breaks-and-societal-costs. Of the US subsidies to the fossil fuel industry, currently 20 percent are allocated to coal and 80 percent to natural gas and crude oil.

6. See Karin Akre, "Dark Triad (Psychology)," *Encyclopaedia Britannica*, last updated February 1, 2025, https://www.britannica.com/science/dark-triad; Jordan Moss and Peter J. O'Connor, "The Dark Triad Traits Predict Authoritarian Political Correctness and Alt-Right Attitudes," *Heliyon* 6, no. 7 (2020): e04453.

7. Upton Sinclair, "I, Candidate for Governor and How I Got Licked," *Oakland Tribune*, December 11, 1934, p. 19, col. 3.

8. Fossil fuel spending to influence US federal elections has increased nearly 67 percent since 1992, with over $200 million donated in the 2023–2024 election cycle. The spending is extremely uneven; the amount given to Democratic candidates has stayed the same, while donations to Republican candidates has quadrupled. See Karin Kirk, "The Fossil Fuel Industry Spent $219 Million to Elect the New U.S. Government," Yale Climate Connections, January 3, 2025, https://yaleclimateconnections.org/2025/01/the-fossil-fuel-industry-spent

-219-million-to-elect-the-new-u-s-government. For influence on the US Supreme Court, see Julia Kane, "The Supreme Court's Climate Decision Came Out of a Decades-Long Campaign to Kneecap Regulation," Grist, June 30, 2022, https://grist.org/accountability/the-supreme -courts-climate-decision-came-out-of-a-decades-long-campaign-to-kneecap-regulation.

9. On cap-and-trade programs' success with acid rain, see US Environmental Protection Agency, "Acid Rain Program," https://www.epa.gov/acidrain/acid-rain-program, accessed April 21, 2025.

10. Environmental activists are not the only targets; so are human rights advocates, journalists, and others. "Close to 2,000 Environmental Activists Killed over Last Decade," Yale Environment 360, September 13, 2023, https://e360.yale.edu/digest/environmental-defenders -murdered-2022; Global Witness, "2016 Land and Environmental Defenders Were Killed Between 2012 and 2023," https://globalwitness.org/en/campaigns/land-and-environmental -defenders/in-numbers-lethal-attacks-against-defenders-since-2012, accessed February 19, 2025.

11. Meenakshi VRai, "The Impact of War on Our Natural Environment," FAWCO, June 27, 2022, https://www.fawco.org/global-issues/environment/environment-articles/4767-impact -of-war-on-our-natural-environment.

12. Joshua Conrad Jackson and Keith Payne, "Cognitive Barriers to Reducing Income Inequality," *Social Psychological and Personality Science* 12, no. 5 (2021): 687–696, https://doi .org/10.1177/1948550620934597; Aurélie Méjean et al., "Climate Change Impacts Increase Economic Inequality: Evidence from a Systematic Literature Review," *Environmental Research Letters* 19 (2024): 043003, https://iopscience.iop.org/article/10.1088/1748-9326/ad376e.

13. Robert Gifford, "The Dragons of Inaction: Psychological Barriers That Limit Climate Change Mitigation and Adaptation," *American Psychologist* 66, no. 4 (2011): 290–302, https://psycnet.apa.org/record/2011-09485-005.

14. The term "code debt," also known as "technical debt," was coined in 1992 by American software developer Ward Cunningham. Vangie Beal, "Technical Debt," *Technopedia*, June 13, 2024, https://www.techopedia.com/definition/27913/technical-debt.

15. Martin Luther King Jr., "Letter from Birmingham Jail," August 1963, https://www .csuchico.edu/iege/_assets/documents/susi-letter-from-birmingham-jail.pdf.

16. Gale M. Sinatra and Barbara K. Hofer, *Science Denial: Why It Happens and What to Do About It* (New York: Oxford University Press, 2021); David Abram, *The Spell of the Sensuous: Perception and Language in a More-Than-Human World* (New York: Pantheon Books, 1996).

17. Brett Jenks, "To Make Progress on Climate Action, Pop 'Normative Bubbles,'" *Behavioral Scientist*, January 30, 2023, https://behavioralscientist.org/to-make-progress-on -climate-action-pop-normative-bubbles.

18. Robert B Cialdini et al., "Managing Social Norms for Persuasive Impact," *Social Influence* 1, no. 1 (2006): 3–15.

19. This is a variant of a quote by H. L. Mencken: "Explanations exist; they have existed for all time; there is always a well-known solution to every human problem—neat, plausible, and wrong." See "Quote Origin: There Is Always a Well-Known Solution to Every Human Problem—Neat, Plausible, and Wrong," Quote Investigator, July 17, 2016, https://quoteinvestigator .com/2016/07/17/solution.

20. "Why Are We Likely to Continue with an Investment Even if It Would Be Rational to Give It Up?," The Decision Lab, https://thedecisionlab.com/biases/the-sunk-cost-fallacy, accessed February 19, 2025.

21. The Dunning-Krueger effect, in which "people who know the least about a topic are often very overconfident of their knowledge, while those who know the most often underestimate their knowledge," is a bias that arises in relation to climate and weather as well. Mark

A. Casteel, "What Do I Know About Severe Weather? The Influence of Weather Knowledge on Protective Action Decisions," *Weather, Climate and Society* 15 (2023): 263–276, https://doi.org/10.1175/WCAS-D-22-0115.1.

22. "The Curse of Knowledge," The Decision Lab, https://thedecisionlab.com/reference-guide/management/curse-of-knowledge, accessed February 19, 2025.

23. K. C. Barr, "Déformation Professionnelle Unveiled: How Your Professional Perspective Could Be Holding You Back," LinkedIn, August 20, 2023, https://www.linkedin.com/pulse/d%C3%A9formation-professionnelle-unveiled-how-your-perspective-k-c-barr.

24. Notions about purity and sanctity and emotions of disgust and anger, often experienced at the gut level, are foundations of human morality and also influence political beliefs. See Jonathan Haidt and Jesse Graham, "When Morality Opposes Justice: Conservatives Have Moral Intuitions That Liberals May Not Recognize," *Social Justice Research* 20, no. 1 (2007): 98–116, DOI: 10.1007/s11211-007-0034-z. These moral responses also arise regarding environmental degradation and those who perpetrate it.

25. I learned a lot about the importance of rigor in scholarship—and the problem of handwaving—from *Ecopsychology* editor Peter Kahn. See his works, including Peter H. Kahn Jr., *Technological Nature: Adaptation and the Future of Human Life* (Cambridge, MA: MIT Press, 2011), and Peter H. Kahn Jr. and Patricia H. Hasbach, eds., *Ecopsychology: Science, Totems, and the Technological Species* (Cambridge, MA: MIT Press, 2012).

26. Framing examples include opposing groups using the same data to advocate for the safety or danger of nuclear power, or for the risks versus benefits of transporting dangerous substances by train. For a survey of factors that affect public opinions about climate issues around the world, see Tien-Ming Lee et al., "Predictors of Public Climate Change Awareness and Risk Perception Around the World," *Nature Climate Change* 5 (2015): 1014–1020, https://doi.org/10.1038/nclimate2728.

Chapter 19: Strategy

1. Min Hyoung Song, *Climate Lyricism* (Durham, NC: Duke University Press, 2022), 15.

2. See writings of science writer Oliver Morton, like *The Planet Remade* (Princeton, NJ: Princeton University Press, 2015) and this essay for a summary: Oliver Morton, "Not-So-Lonely Planet," *New York Times*, December 23, 2008, https://www.nytimes.com/2008/12/24/opinion/24morton.html.

3. Mitchell Thomashow, *Ecological Identity: Becoming a Reflective Environmentalist* (Cambridge, MA: MIT Press, 1996), 153–154.

4. Flights may be necessary for health or your family. You can cancel out your emissions by supporting an offset project that draws down carbon in another location, or create your own offset by lowering emissions or doing service to the environment in other parts of your life. See Angelo Gurgel, "Explainer: Carbon Offsets," Climate Portal, Massachusetts Institute of Technology, last updated November 8, 2022, https://climate.mit.edu/explainers/carbon-offsets.

5. "Scope 1, 2, and 3 Emissions Explained | Climate Now Mini Video 01," YouTube, posted by Climate Now, April 13, 2022, https://www.youtube.com/watch?v=ck3uW0bA78k.

6. The UN Sustainable Development Goals are described at https://sdgs.un.org/goals, accessed April 21, 2025.

7. See "What Are Sanctions?," Comply Advantage, last updated January 22, 2025, https://complyadvantage.com/insights/what-are-sanctions. On regulation of ozone-depleting CFCs, see "Ozone-Depleting Substances on the Black Market," United States Environmental Protection

Agency, last updated October 7, 2024, https://www.epa.gov/ozone-layer-protection/ozone
-depleting-substances-black-market. On combating illegal trade in wildlife, see "Illegal Wildlife
Trade: Enhancing Responses to Wildlife Crime and Illegal Trade," TRAFFIC, https://www.traffic
.org/about-us/illegal-wildlife-trade, accessed February 19, 2025. On stopping human trafficking, see Lindsey King, "International Law and Human Trafficking," *Human Rights and Human
Welfare* 9, no. 1 (2009): 32, https://digitalcommons.du.edu/hrhw/vol9/iss1/32.

8. "Volkswagen Violations," United States Environmental Protection Agency, last updated
October 9, 2024, https://www.epa.gov/enforcement/learn-about-volkswagen-violations.

9. A common approach to conservation and sustainability is called Community Based
Social Marketing, created by Doug McKenzie-Mohr; see https://cbsm.com/about, accessed
February 19, 2025.

10. See *80,000 Hours: Find a Fulfilling Career That Does Good* (Oxford: Trojan House,
2023), and website, https://80000hours.org/start-here.

11. I learned this from longtime Portland, Oregon, environmental leaders Dick and
Jeanne Roy, founders of the Northwest Earth Institute (now known as Ecochallenge.org) and
the Center for Earth Leadership.

12. Hannah Ritchie, *Not the End of the World: How We Can Be the First Generation to
Build a Sustainable Planet* (New York: Little, Brown Spark, 2024), 297.

13. On having an exit strategy, see *80,000 Hours* podcast #207, Sarah Eustis Guthrie on why
she shut down her charity, and why more founders should follow her lead: https://80000hours
.org/podcast/episodes/sarah-eustis-guthrie-founding-shutting-down-charity.

14. The full quote from Bill Gates: "The first rule of any technology used in a business
is that automation applied to an efficient operation will magnify the efficiency. The second
is that automation applied to an inefficient operation will magnify the inefficiency," https://
www.azquotes.com/quote/107323. For a perspective on climate mitigation, see Bill Gates,
How to Avoid a Climate Disaster (New York: Knopf, 2021).

15. On design thinking, see, for example, Raz Godelnik, "Design for the Climate Crisis—
An Updated Approach," Medium, August 19, 2024, https://razgo.medium.com/design-for-the
-climate-crisis-an-updated-approach-d74f2b318c95.

16. One of many online versions of the classic Oblique Strategies card deck created by
Brian Eno and Peter Schmidt is at https://ob-strat.netlify.app. See also Brian Eno and Bette
Adriaanse, *What Art Does: An Unfinished Theory* (London: Faber & Faber, 2025).

17. Randy Olson, *Houston, We Have a Narrative: Why Science Needs Story* (Chicago:
University of Chicago Press, 2015); Randy Olson, *Don't Be Such a Scientist*, 2nd ed. (Washington, DC: Island Press, 2018).

Chapter 20: Weathering a Disaster

1. "Sin-Killer Griffin—Wasn't That a Mighty Storm," YouTube, posted by The Soul of
Sam Collins, October 23, 2012, https://www.youtube.com/watch?v=2XHOi-hG3Y0; "Wasn't
That a Mighty Storm," Library of Congress, https://www.loc.gov/item/afc9999005.593.

2. "'Titanic Will Founder'—Scene HD," YouTube, posted by Titanic Movie/Pelicula,
August 16, 2018, https://www.youtube.com/watch?v=SP7BWb1ndpA.

3. In the language of insurance and risk management, this makes it difficult to determine the extent of the peril (risks we can prevent), be clear on the hazards (risks we cannot
know or prevent), or quantify the likelihood of harm (the actual damages) that will occur.
Put simply, you cannot insure something if you can't quantify the risks and what the cost of
the harm would be.

4. The *Titanic* story is unfortunately an analogy for climate change and the lack of societal leadership that could have prevented the worst of what we are experiencing now. Nathaniel Rich, *Losing Earth* (New York: Picador, 2020); John Lanchester, "Two New Books Dramatically Capture the Climate Change Crisis," *New York Times*, April 12, 2019, https://www.nytimes.com/2019/04/12/books/review/david-wallace-wells-uninhabitable-earth-nathaniel-rich-losing-earth.html.

5. Adam Aton and E&E News, "Successive Disasters Put U.S. Gulf States at Risk of a 'Knock-Out Blow,'" *Scientific American*, July 18, 2024, https://www.scientificamerican.com/article/u-s-gulf-states-risk-knock-out-blow-from-successive-disasters.

6. The concept of disaster capitalism has been described by Naomi Klein, among others. See Naomi Klein, *The Shock Doctrine: The Rise of Disaster Capitalism* (New York: Henry Holt, 2007). For a discussion of hurricane recovery in Puerto Rico, see Yarimar Bonilla, "The Coloniality of Disaster: Race, Empire, and the Temporal Logics of Emergency in Puerto Rico, USA," *Political Geography* 78 (2020): 102181, https://doi.org/10.1016/j.polgeo.2020.102181.

7. Ryan Hass, "Vigilante Activity Persists After Oregon Fires, Leads to Citations," Oregon Public Broadcasting, September 18, 2020, https://www.opb.org/article/2020/09/18/antifa-rumors-multnomah-county-oregon-wildfires.

8. Storm surges are immense mounds of water built up by ocean storms that can advance miles on land and travel hundreds of miles up rivers. Because Galveston was essentially constructed on a sandbar, a storm surge from the 1900 hurricane caused widespread death and destruction there. See "Galveston Hurricane of 1900," National Park Service, https://www.nps.gov/articles/galveston-hurricane-of-1900.htm, accessed February 28, 2025.

9. National Weather Service, "Hurricane Ike Storm Surge Inundation Maps," https://www.weather.gov/lch/ikesurge, and "Hurricane Harvey and Its Impacts on Southeast Texas (August 25–29, 2017)," https://www.weather.gov/hgx/hurricaneharvey, both accessed February 12, 2025.

10. The Houston area contains critical infrastructure for the US energy, chemical, transportation, and critical manufacturing sectors. See Neena Satija, Kiah Collier, Al Shaw, and Jeff Larson, "Hell and High Water," *Houston Tribune* and ProPublica, March 3, 2016, https://houston.texastribune.org/hell-and-high-water.

11. For comparison, the *Exxon Valdez* oil spill in 1989 released about 12 million gallons of crude oil into the open waters off the Alaskan coast; the Deepwater Horizon explosion in 2010 released about 210 million gallons of crude oil into the Gulf of Mexico.

12. Examples of volunteer disaster relief programs include faith-based groups like Texans on Mission, https://www.texansonmission.org/disaster-relief, and the Massachusetts-based All Hands All Hearts, https://www.allhandsandhearts.org/about-us, both accessed February 11, 2025.

13. E. Packard, "Grit: It's What Separates the Best from the Merely Good," *Monitor on Psychology* 38, no. 10 (2007): 10, https://www.apa.org/monitor/nov07/grit.

14. Stevan E. Hobfoll et al., "Five Essential Elements of Immediate and Mid-Term Mass Trauma Intervention: Empirical Evidence," *Psychiatry* 70, no. 4 (2007): 283–315, 316–369.

15. The resource site created after the 2023 Maui wildfire is American Academy of Pediatrics, Hawaii Chapter, "Maui Resources and Information After the Fires," https://www.aaphawaii.org/maui-resources, accessed February 11, 2025. Information on how to coordinate volunteers during and after disasters can be found on the Oregon Serves website at https://www.oregon.gov/oregonserves/emergency-response/pages/disaster-assistance-resources.aspx, accessed February 11, 2025.

16. Kathryn Schultz, "What a Major Solar Storm Could Do to Our Planet," *New Yorker*, February 26, 2024, https://www.newyorker.com/magazine/2024/03/04/what-a-major-solar-storm-could-do-to-our-planet.

17. Kathryn Schultz, "The Really Big One," *New Yorker*, July 20, 2015, https://www .newyorker.com/magazine/2015/07/20/the-really-big-one.

Epilogue

1. From the Tao Te Ching, chapter 3. This is my translation based on a literal reading and inspired by the Ursula Le Guin and Stephen Mitchell interpretations I am most familiar with. Word for Word Translation, Center Tao: https://www.centertao.org/essays/tao -te-ching/carl.

2. For example, the recent Lord of the Rings film trilogy omitted the crucial return and reclaiming of the Shire by the hobbits. Michael John Petty, "'The Return of the King' Ending Most Lord of the Rings Adaptations Ignore," Collider, February 25, 2024, https://collider .com/lord-of-the-rings-original-ending-explained. For a deeper Tolkien dive: "Scouring of the Shire," The Lord of the Rings Wiki, https://lotr.fandom.com/wiki/Scouring_of_the _Shire#, accessed April 21, 2025.

3. From the work of twentieth-century mythologist Joseph Campbell. Michael Langdon, "Master of Two Worlds and the Hero's Journey," in *Encyclopedia of Heroism Studies* (Cham, Switzerland: Springer, 2024), https://doi.org/10.1007/978-3-031-17125-3_320-1. For feminist and women-centered readings, see Maureen Murdock, *The Heroine's Journey* (Boston: Shambhala, 1990), and Victoria Lynn Schmidt, *45 Master Characters: Mythic Models for Creating Original Characters* (Cincinnati: Writer's Digest Books, 2001). For Murdock, the final stage permits the heroine to see through binaries of masculine and feminine and to interact with a complex world that includes the heroine but is also larger than their personal lifetime or their geographical / cultural milieu. For Schmidt, the final stage is called "Return to a World Seen with New Eyes."

INDEX

Note: Italic page numbers refer to illustrations.

ABOUT THE AUTHOR

DR. THOMAS DOHERTY IS AN AWARD-WINNING CLINICAL PSYCHOLO-
GIST, internationally recognized for his research on the psychological impacts
of climate change. His practice, Sustainable Self, marries traditional therapeutic approaches with findings about sustainability, nature, and well-being
under an ecopsychology framework, and he has trained and supported mental health professionals from around the world addressing eco-anxiety.

Doherty is a regular guest speaker at conferences and panels worldwide, and has conducted workshops and trainings for organizations such
as the US National Park Service, the California Department of Health, the
Association of Zoos and Aquariums, and the Aspen Ideas Festival. A Fellow of the American Psychological Association, Dr. Doherty is the creator of the podcast *Climate Change and Happiness*, which he cohosts with
researcher Panu Pihkala. He lives in Portland, Oregon.